Fate and Prediction
of
Environmental Chemicals
in
Soils, Plants, and
Aquatic Systems

Fate and Prediction
of
Environmental Chemicals
in
Soils, Plants, and
Aquatic Systems

Edited by
Mohammed Mansour

CRC Press
Taylor & Francis Group
Boca Raton London New York

CRC Press is an imprint of the
Taylor & Francis Group, an **informa** business

ngress Cataloging-in-Publication Data

iction of environmental chemicals in soils, plants, and
stems / edited by Mohammed Mansour.
      cm.
ibliographical references and index.
371-616-7 (acid-free paper)
icides--Environmental aspects.   2. Soil pollution.
Jnderground--Pollution. 4. Plants, Effect of pesticides
isour, Mohammed.
7F38   1993
c20
                                                      92-42304

Congress record exists under LC control number: 92042304

1-315-89290-0 (hbk)

PREFACE

We gave these proceedings the title "Study and Prediction of Pesticides Behaviour in Soils, Plants, and Aquatic Systems" because we want to show our efforts in identifying, interpreting, and predicting environmental impact of different substance classes.

The workshop attempted to review concepts and ideas regarding environmental development and to examine the distribution and effects of these substance classes in different environmental compartments. Therefore the meeting aimed to link knowledge of chemistry with the environmental problems to be solved by chemical methods. Any pollution control action today must take into consideration not only the soil contamination and water problems but also air pollution. These proceedings are intended to provide guidance to scientists and researchers as well as people involved in regulatory action, monitoring work, and estimating the hazard of chemical substances in ecosystems. It is hoped that these topics will catalyze further research on the persistence, toxicity, distribution, accumulation, metabolization, and transformation of pesticides and other environmental chemicals of similar structures and properties. This third workshop attracted participants from different countries and the discussions were intense. The participants had differing backgrounds and skills, all of them relevant to aspects of the meeting topics.

As organizer of the workshop I wish to take this opportunity to express my thanks to GSF-Forschungszentrum für Umwelt und Gesundheit GmbH München/ Neuherberg and to the Congreß-Dienst for the invaluable assistance rendered in the preparatory work for the meeting as well as their cordial hospitality and assistance during the event in Neuherberg.

I would like to express my personal gratitude to the many scientists of different countries who attended the workshop and contributed to its success with their presentations, posters, and discussions. Finally, I hope that all the information in these procedings will stimulate further research in such an exciting field.

Mohammed Mansour

THE EDITOR

Mohammed Mansour, Ph.D., is a research chemist at the Center for Environment and Health [G.S.F.] Neuherberg-Munich, Germany. Dr. Mansour was born in 1942 in Casablanca, Morocco. He received a degree in Chemistry (1973) and Ph.D. in the field of Organic Chemistry and Pharmacology (1975), both from the University of Bonn. He spent a few months at the University of La Sapienza in Rome in the Department of Chemistry as a guest Professor. He is the author or co-author of more than 130 publications. He serves on the Editorial Advisory  Boards of Fresenius Environmental Bulletin. His research interests are the fate and transport of organic compounds in soil and water. Current areas include the development of techniques concerning the photo process formation of pesticides on soil surfaces. He has organized and chaired or directed a series of ecotoxicological symposia since 1986. He conducted International Cooperation and he received the 1986 International award for research in the field of environmental protection from the Technical University of Lublin in Poland.

CONTENTS

CHAPTER 1

Transport and Transformation of Pesticides in Soil

Irene Scheunert

I. INTRODUCTION

In most cases, pesticides are xenobiotic organic substances which are applied intentionally to terrestrial ecosystems. However, when they have accomplished their intended effects, their residues which are left in the ecosystem must be regarded as undesirable environmental chemicals. The behavior of these residues in the soil-plant system is governed by the same rules as that of other organic xenobiotic pollutants for which input into the soil was not intended by man.

After the input of pesticides into the soil system, various physical and physicochemical processes, as well as chemical and biochemical reactions, determine their residue behavior. Physical and physicochemical processes govern the transport within and out of the soil system. Chemical and biochemical reactions effect transformations of the pesticide, resulting in molecular alterations and in degradation down to complete mineralization.

Alterations in the chemical structure of a pesticide, as performed by biotic or abiotic conversion reactions, lead to the formation of new xenobiotic compounds. Due to the changes of molecular substituents, these compounds may differ significantly from the parent compounds in physicochemical properties and also in their mobility and distribution within the system. These facts result in further complications in prediction of the fate of a pesticide in the soil-plant system and thus in evaluation of its ecological significance.

1

In this chapter, an overview of the information available today both on transport and on transformation and degradation of pesticides in soil is given, and possibilities of prediction are discussed.

II. TRANSPORT OF PESTICIDES WITHIN SOIL AND TRANSFER FROM SOIL TO OTHER MEDIA

A. Adsorption-Desorption Processes in Soil

Adsorption-desorption processes in soil play a paramount part in all physical processes affecting the residue behavior of pesticides in the soil-plant system, such as volatilization from soil into the air, soil-water transfer resulting in mobility in soil and in leaching of pesticides to surface- or groundwater, and uptake by plants or soil fauna. Additionally, biological activity and degradation also are affected by adsorption.[1] In general, adsorption is defined[2] as the adhesion or attraction of one or more ionic or molecular layers to a surface. In soil, the sum of all kinds of fixation of ions or molecules on or within the solid phase is called adsorption.

Pesticides may be adsorbed by soil organic matter as well as by inorganic soil fractions, and the relation between adsorption at both sites depends on soil properties as well as chemical structure of the pesticide. Adsorption by inorganic and organic matter is not by just one mechanism of pesticide-soil interaction. Physical, physicochemical, and chemical processes are involved. There is a continuum of possible adsorption interactions, starting with fixed site adsorption and ending with partition between three-dimension phases such as aqueous solution and soil organic matter.[3] Bonds vary between reversible ones and those resulting in so-called bound residues.

The adsorption of pesticides may be quantitatively characterized by the linear adsorption coefficient K_D[4] or by the adsorption constants derived either from the Langmuir or the Freundlich adsorption equations.[5] The linear adsorption coefficient K_D is the quotient of the concentration in the solid soil phase divided by that in the liquid soil phase. It is applicable to characterizing the adsorption from dilute aqueous solutions where the partition between solid and liquid soil phases is more or less independent on concentration.

Numerous publications have demonstrated that adsorption is significantly correlated with soil organic carbon, at least in the case of nonionic pesticides. Therefore, the use of adsorption coefficients related to soil organic carbon only may be appropriate. The adsorption coefficient related to soil organic carbon, K_{OC}, is K_D multiplied by 100 and divided by the percent of organic carbon in soil.

The linear adsorption coefficient K_D is sometimes not sufficiently exact to describe the adsorption of a pesticide over a larger range of concentrations. The relations between the concentrations in the liquid and the solid phase can be described by isotherms. The Freundlich isotherm is derived on the assumption that a decrease in heat of adsorption with increasing surface concentration or coverage can occur due to surface heterogeneity. The Freundlich isotherm equation is written as follows:

$$\frac{x}{m} = K_F \times C_e^{1/n}$$

where   x/m = pesticide concentration in the solid phase
        C_e = concentration in the liquid phase at equilibrium
        K_F = adsorption constant
        1/n = another constant providing a rough estimate of the intensity
           of adsorption.

The desorption of adsorbed pesticides by water or by electrolytes in most cases is not complete, i.e., adsorption is not fully reversible. The nondesorbable part of pesticides is due either to strong bonds which are broken only by solvent extraction, or to even stronger bonds which result in the formation of "bound residues."

Adsorption coefficients and constants related to soil organic carbon have been shown to be correlated with simpler physicochemical parameters. Empirical equations have been obtained experimentally for nonionic chemicals relating soil adsorption to water solubility, n-octanol-water partition coefficients, parachor, and molecular connectivity indices. Both molecular size and structure are important in adsorption of a chemical compound by the soil. Dissociation constants play a role for ionizable pesticides.[6] The n-octanol-water partition coefficient is the preferred physicochemical parameter to predict soil adsorption. However, it should be considered that this parameter is correlated only with that part of the adsorbed pesticide which is reversibly fixed by "partition" in the lipophilic organic soil fraction.

B. Mobility of Pesticides in Soil

Mobility of pesticides in soil is related to adsorption, on the one hand, and the mass flux of dissolved fractions, on the other hand. Adsorption retards the mass flux which consists of diffusion, convection, and dispersion and which also is a function of the removal of pesticides from the solution by biological and chemical reactions.[7]

Diffusion (diffusive mass flux) is a physical process by which molecules, atoms, and ions — due to their thermic mobility — move from sites of higher concentration to those of lower concentration. This transport mechanism is independent of water movement.

Convection (convective mass flux) is the passive movement of solutes with the moving water. It is a function of the volume flow velocity of the water.

Dispersion (dispersive mass flux) is the distribution or mixing of solutes in the moving pore water, which results from different flow velocities of individual water volumes. Like convection, it represents a current-dependent contribution to mass flux in soil water.

According to these processes affecting pesticide mobility in soil, relevant soil parameters (apart from adsorption coefficients and microflora performing biodegradation) are soil texture, porosity, density, moisture ratio in the layers, water flow velocities, and aggregate stability.[1]

C. Transfer of Pesticides from Soil to Water

Surface run-off, tile drainage, and percolation into deeper groundwater are the main processes of water — and thus pesticides — leaving the root zone. Consequently, these processes are potential sources of contamination of rivers, lakes, and groundwater with pesticides.

The contamination of groundwater and potable water with pesticides has been a source of major concern for the last few years. Therefore, the prediction of leaching rates of pesticides from soil is becoming more important for assessing possible concentrations in groundwater and potable water for the future. In mathematical models under discussion, the most sensitive parameters are the adsorption coefficient K_{OC}, the Freundlich exponent $1/n$, and the half-life of degradation (DT_{50}). The variability of K_{OC} between different soils is a source of uncertainty in groundwater contamination prediction. Thus, a variation of K_{OC} by a factor of 5, which is quite realistic, changes the predicted pesticide concentration in groundwater by at least one order of magnitude.[8,9] The relationship between K_{OC} and predicted pesticide concentration in groundwater is also affected by variations in the Freundlich exponent between 0.8 and 1.0. The DT_{50} of biological degradation is also uncertain since degradation of pesticides in soil is neither linear, nor does it follow first-order kinetics in most cases; and degradation kinetics are widely unexplored. The reduction of microbial concentrations in the soil below the surface horizon also has to be taken into account.[1]

In spite of the outstanding importance of adsorption properties and degradation rates for the leaching of pesticides from soil, soil texture may be decisive in special cases. Organic matter in soil affects the mobility and leaching of pesticides not only negatively by adsorption, but also positively by creating a higher aggregate stability. Furthermore, organic matter has a positive effect on pesticide mobility and leaching by the formation of water-soluble adducts between poorly water-soluble pesticides and readily water-soluble humic substances.[10]

Temperature also affects leaching of pesticides in soil. Interactions between different organic chemicals concerning mobility and leaching in soil may occur through mutual influences on their water solubility or by competition for adsorption sites in the soil solid phase.

The interaction of all parameters discussed above and of further unexplored ones may result in the appearance of pesticide residues in groundwater. Trace amounts of pesticides detected in groundwater after agricultural use[11] include those for which penetration into groundwater was not predicted by theoretical and experimental models. In former years, chlorinated pesticides such as aldrin and dichlorodiphenyltrichloroethane (DDT) were the pesticides preferably found in potable water.[12] Their presence was not predicted, due to their low water solubility; however, by their biological and chemical persistence they were able to reach groundwater. Today, pesticides detected in groundwater after agricultural use belong to various chemical classes, including carbamates, triazines, acetanilides, uracils, dinitrophenols, and others.[11] Traces of readily biodegradable pesticides may be found in groundwater at analytically detectable levels if their mobility is high.

D. Transfer of Pesticides from Soil to the Atmosphere

Volatilization of pesticides from soil is the transfer of the pesticides as a gas through the soil-air interface under environmental conditions. Volatilization of pesticides from soil has been underestimated for a long time. Since pesticides, in general, have low vapor pressures, volatilization has been regarded as being of little importance, both for residue decline in soil and for potential air contamination by pesticide vapors. In many cases, volatilization was misinterpreted as degradation.

However, recent research results have demonstrated that volatilization plays an important role in the fate of pesticide residues in terrestrial ecosystems as well as in the total environment. For persistent pesticides with low degradation rates in soil, volatilization is the main pathway of residue decline in the terrestrial ecosystem. On the other hand, volatilization of pesticides may contribute significantly to air pollution and thus to the long-range transport of pesticides by the air.

For the volatilization from dry soil surfaces, the vapor pressure of the adsorbed pesticide as well as the gas-phase transfer velocity depending both on pesticide molecular properties and environmental parameters such as air velocity are important. In the lower parts per million range, the vapor pressure of adsorbed pesticides is significantly lower than that of nonadsorbed substances; with increasing concentration of the pesticide, the vapor pressure rises and reaches that of the nonadsorbed compound at a concentration level which is dependent both on properties of the pesticide and those of the soil.[13] Thus, the importance of soil adsorption coefficients and of soil organic matter in volatilization processes is evident.

Volatilization of pesticides from moist soil surfaces probably is more complex than that from dry soil. It is assumed that volatilization takes place from the soil aqueous phase after desorption of the pesticide from the solid soil phase.[14] In addition to the factors relevant for volatilization from dry surfaces, the Henry constant H is important for volatilization from moist surfaces. In general, volatilization from moist surfaces is faster than from dry ones.

For the volatilization of pesticides incorporated in soil, the transport from deeper soil layers to the evaporating surface is a very important factor. Pesticides with lower Henry's constants (e.g., prometon) accumulate at the soil surface with water evaporating, and volatilization increases with time; volatilization is controlled by the air boundary-layer thickness above the soil surface, by water evaporation rate, and by H. Pesticides with higher Henry's constants (e.g., lindane) do not accumulate at the soil surface, and volatilization decreases with time (with or without water evaporating); the transport from the soil to the evaporating surface is the limiting factor in the total volatilization process.[15]

Sources other than volatilization in the vapor state may contribute to atmospheric contamination with pesticides. These are, e.g., the whirling up of sprays and the suspension of soil particles in air. In the case of low volatile pesticides such as atrazine, it was demonstrated that atmospheric transport occurs nearly exclusively by the particle phase that is not in equilibrium with the gas phase.[16] This suggests that emission as aerosols prevails over that as a vapor due to volatilization.[16]

Resulting from volatilization and the other emission sources discussed, a wide range of pesticides has been detected in air, fog, water, and rain comprising organochlorine and organophosphorous insecticides, as well as various herbicides.[16-18] This demonstrates that potential air contamination has to be considered for every pesticide, regardless of its vapor pressure.

III. TRANSFORMATION AND DEGRADATION OF PESTICIDES

In soil, pesticides undergo various abiotic and biotic transformation reactions.

Abiotic reactions comprise all of those reactions which are not enzymatic but initiated by reactive chemical species or molecular functions in soil, or by catalysis by nonliving soil constituents such as metal oxides and organic or mineral surfaces. On the contrary, biotic reactions are catalyzed by enzymes. Up to now, it is still very difficult to distinguish between abiotic and biotic causes of chemical reactions in soil. In many cases, the same reaction product partly originates from an enzymatic and partly from an abiotic process. The transition from one mechanism to the other is neither abrupt nor well-defined. The differentiation between both mechanisms by using sterilized and nonsterilized soil is not indicative of the proportion between abiotic and biotic reactions occurring in natural soil, since sterilization — besides eliminating enzymatic activity — produces changes both in soil structure and in content of abiotic reactive species such as free radicals in soil or singlet oxygen in the water phase. Therefore, the abiotic portion of conversion products obtained by sterilized soils should be regarded as the minimum limit of potential abiotic reactions.

Table 1. Abiotic Reactions of Pesticides in Soil

Type of transformation	Reaction	Examples of pesticides
Redox reactions	Oxidation	Arsenious acid, organomercury pesticides, anilines, phorate
	Oxidative coupling	Phenols, anilines
	Mineralization	Amitrole
	Reduction	Trifluralin
	Reductive dechlorination	DDT, Mirex, toxaphene
Reactions catalyzed by clay surfaces	Hydrolysis	Organophosphates, phosmet, ronnel, s-triazines
	Ester formation	Alcohols, alkenes
	Rearrangements	Organophosphates, ronnel
	Oligomerization/polymerization	Olefines, dienes
	Dehydrochlorination	DDT
Photolysis	Direct photolysis	Fluchloralin, bentazone, flumetralin
	Reaction with singlet oxygen	Disulfoton
	Reaction with ozone	Parathion
Interactions between different pesticides	Direct reaction	Degradation products of metham-sodium, methamsodium ethylene dibromide
	Promotion of reactions	Atrazine/fenaminosulf

Source: Scheunert[19] and Wolfe et al.[21]

Attempts to differentiate between chemical and microbial transformations by their different activation energies (microbiological: 4–6 kcal/mol; chemical: 18–25 kcal/mol) also failed because some enzyme systems were shown to have activation energies which fall into the range of chemical degradation processes (bacterial dehydrogenase: 15–25 kcal/mol). However, the question as to whether a given pesticide is degraded by microbial or chemical mechanisms can be considered of more academic than of practical interest and certainly is not important from an environmental safety point of view.[19,20]

A. Abiotic Reactions

Table 1 gives some examples of abiotic reactions of pesticides in soil.

Opposite to most of the oxidations, reductions, and hydrolyses which initiate degradation or at least do not imply the synthesis of larger molecules, reactions between pesticides and natural soil constituents or between two different pesticides or pesticide degradation products result in molecule enlargement. If the reaction products between xenobiotics and natural soil constituents are of low molecular weight, they are, in general, water soluble. Such adducts may be responsible for the transport of normally insoluble pesticide residues in leaching water. If the reaction partner is a humic acid precursor in soil, polymerization continues and leads to higher molecular products. These are insoluble and are part of the soil-bound pesticide fraction discussed below.

It is assumed that humic substances are formed by the polymerization of phenols liberated by the decomposition of plant material, especially lignin, or synthesized by microorganisms. In particular, o- and p-diphenols are polymerized, the polymerization being accompanied by oxidation of these phenols which are highly sensitive to oxidation.[22] Amino acids can be involved in these reactions in two ways: one is addition, and the other is deamination and decarboxylation of the amino acid. The participating natural amino compounds may be replaced by amines of xenobiotic origin, such as chlorinated anilines which are degradation products of various herbicides in soil. A lower molecular product of oxidative coupling between catechol and 4-chloroaniline has been identified by Adrian et al.[23] as an anilinoquinone. The reaction is catalyzed by pyrolusite, but also enzymatically. According to the natural humification process, this reaction proceeds to larger polymeric structures containing the xenobiotic substance as a covalently bound residue.

Reactions between the components of pesticide mixtures in soil, resulting in the formation of new foreign compounds, are not frequent since normally the concentrations are too low and the reactivity of pesticides, in general, is not sufficient. However, a few examples have been reported for pesticides in soils, as shown in Table 1.

The fumigant metham-sodium (Vapam) is decomposed in soil, resulting in a mixture of chemicals with components that may react with each other to form a new chemical.[24] The concentration of active metham-sodium applied to soil was about 500 mg/kg.

The same pesticide was shown to react in soil with halogenated hydrocarbons contained in certain nematocides. Metham-sodium was alkylated at its sulfide group by ethylene dibromide; 1,2-dibromo-3-chloropropane; 1,3-dichloropropene; and related chlorinated hydrocarbons.[25] The concentration range in soil was about 40–90 mg/kg.

The fungicide Fenaminosulf or one of its degradation products promotes the conversion of atrazine to its hydroxy derivative.[26] The observed Fenaminosulf antagonism toward herbicidal activity of s-triazines probably is due to this phenomenon for which the mechanism is not known.[27]

These examples show that, in principle, such reactions are possible if the concentrations in soil are sufficiently high and the pesticides or their degradation products have substituent groups of sufficient reactivity.

B. Biotic Reactions

Biotic transformations in soil are those occurring in living soil organisms or catalyzed by enzymes within or outside of cells.

In the biotic attack on pesticides, two possibilities have to be considered. In the first case, the pesticide is degraded to low molecular inorganic products such as carbon dioxide, water, chloride ions, etc. (mineralization) or to low molecular organic fragments joining the natural carbon pool. This degradation (metabolism)

Table 2. Biotic Reactions of Pesticides in Soil

Type of transformation	Reaction	Examples of pesticides
Oxidative processes	C-Hydroxylation	Cypermethrin, carbofuran, aromatic compounds
	β-Oxidation	Phenoxy alkanoic acids
	Epoxidation	Cyclodienes
	Ketone formation	Carbofuran
	C=C cleavage	Cyclodienes, aromatic compounds
	C-dehydrogenation	Lindane
	N-oxidation	Anilines
	N-demethylation	Phenylureas
	S-oxidation	S-containing pesticides
	Substitution of O for S	Organophosphorous compounds
	Ether cleavage	Phenoxy alkanoic acids
	Mineralization	Nearly all organic compounds
Reductive processes	C-reduction	Alkenes and alkines
	N-reduction	Nitro compounds
	S-reduction	Sulfoxides, disulfides
Hydrolytic processes	Ester hydrolysis	Carboxylic esters, sulfates
	Carbamate hydrolysis	Carbamate insecticides
	Nitrile hydrolysis	2,6-Dichlorobenzo-nitrile, cypermethrin
	Epoxide hydrolysis	Dieldrin, intermediates of aromatic C-hydroxylation
Synthetic processes	Conjugation	Butachlor, dichlorophenol
	Reaction with natural humic monomers	Chlorinated anilines and phenols
	Interaction between different pesticides	Chlorinated anilines and phenols

Source: Scheunert[19] and Mansour et al.[28]

is accompanied by a growth of biomass of the respective organisms, demonstrating that they obtain from this reaction both energy and carbon for biosynthesis. This kind of biological reaction, however, is confined to a limited number of pesticides and to a few microorganism strains which are able to grow on these pesticides as a sole carbon source. The second mechanism is called cometabolism. Microorganisms bring about chemical transformation of pesticides, but obtain neither carbon nor energy from the reaction which, consequently, does not promote growth.

Biotic transformation products (metabolites) of pesticides found in soil originate primarily from cometabolic processes; however, they may also represent intermediates in a mineralization sequence. Some biotic reactions of pesticides occurring in soil are presented in Table 2.[19,28] Degradative processes comprise oxidative, reductive, and hydrolytic processes. It should be emphasized, however, that there exist a number of microbiological reactions which cannot be classified in this scheme, such as isomerizations, molecular rearrangements, or loss or addition of various substituents. In many cases, the reaction mechanism is not known; and sometimes conversion products regarded as resulting from hydrolytic attack are actually produced by mixed function oxidases, as in some cases of organophosphorus compounds.[29]

Table 3. Biotic Dechlorination of Pesticides in Soil

Mechanism	Reaction	Examples of pesticides	Ref.
Oxygenolytic dechlorination	Introduction of oxygen from the air	2-Chlorobenzoate, 4-chlorophenyl-acetate	33
Reductive dechlorination	Substitution of H for Cl	DDT, dieldrin, pentachlorophenol	34–37
	Removal of Cl by formation of C=C bonds	Lindane	38–40
Hydrolytic dechlorination	Introduction of oxygen from water	Trichloroacetic acid	41
		Chloroallylalcohols	42
		Pentachlorophenol	43–44
		Atrazine	45
Dehydrochlorination	Removal of HCl by formation of C=C bonds	DDT	46
		Lindane	47
Dechlorination by conjugation	Substitution of S–CH$_3$ for Cl	Pentachloronitrobenzene (earthworms)	48

Like abiotic processes, synthetic reactions of pesticides may occur also as biotic processes. Conjugations with larger natural molecules such as carbohydrates or amino acids, resulting in an enhancement of water solubility, are much less frequent in soil microorganisms than in higher plants or animals. Reactions of pesticides or their primary conversion products with natural humic monomers may occur as biotic processes also; the oxidative coupling reaction between catechol and 4-chloroaniline, as described in the paragraph on abiotic reactions, is catalyzed not only by metal oxides in soil but also by oxidizing enzymes.[23] Xenobiotic phenols such as chlorinated phenols themselves may take the role of humic acid precursors and react with themselves to form oligomers, the reaction being the beginning of the formation of "chlorinated xenobiotic" humic acids which also may be regarded as soil-bound pesticide residues.

Interactions between different pesticides, as catalyzed by enzymes, have been observed for different chlorinated anilines which are metabolites of various phenylurea herbicides in soil. They may be dimerized to mixed azobenzenes.[30,31] Cross-coupling between chlorinated phenols and chlorinated anilines results in trimeric products.[32]

Dechlorination of chlorinated pesticides, the cleavage of the C– Cl bond, is a highly important step in the degradation of this xenobiotic substance class. Because of their ecotoxicological importance, dechlorination processes occurring in soil will be discussed in a more detailed way.

Although the C– Cl bond is not absolutely foreign in nature but is synthesized by some living organisms, it is, nevertheless, the main reason for the general slow degradability of organochlorine compounds and their persistence in the environment. Dechlorination is achieved by either oxidative, reductive, or hydrolytic mechanisms; or by conjugation followed by elimination of the conjugating molecule or a part of it. The most important mechanisms are shown in Table 3; examples are presented also.

In all of these cases, dechlorination initiates the degradation of a number of persistent pesticides by splitting off the most stable substituent in the molecule.

Even if the degradation is not continued in the same species, the dechlorinated product may be attacked by other species or abiotically.

Aromatic rings with catechol structure are degraded either by o-scission between both hydroxyl groups or by m-scission beneath them, resulting in carbon dioxide and in small fragments undergoing normal metabolism.[49] The so-called homogentisin pathway is a natural way to degrade, e.g., aromatic amino acids.[50] Xenobiotic substances may follow this degradation pathway if homogentisin is formed by preceding transformation processes. Biotic degradation of DDT by the bacterium *Pseudomonas aeruginosa* is an example where a metabolic sequence under various environmental conditions runs into this compound for which the ring then is opened and metabolized further, resulting in fumaric acid and acetylacetic acid. One should, however, not conclude that DDT is a readily biodegradable compound. The metabolic steps leading to homogentisin require varying cosubstrates and aeration conditions, i.e., a complex gradient of environmental conditions that is difficult to maintain in the laboratory and that is nearly impossible in the natural environment.[34]

As an example of various transformation pathways of a pesticide, the conversion of the fungicide pentachloronitrobenzene (quintozen) will be presented (Figure 1).

The main modifications in the parent molecule are those of the nitro group which is either reduced to an amino group, or replaced by hydrogen, or by a hydroxyl group which is then methylated, or by a sulfhydryl group which is then methylated and oxidized. The dechlorinated products shown in Figure 1 were not isolated from soil, but from plants which probably had taken them up from soil.[51]

By systematic studies of the transformation potential of pesticides through their chemical molecular characteristics, the prediction of the conversion of pesticides becomes possible by predicting the conversion of their substituents. In addition, substituents or molecular moieties not directly involved in the metabolic reactions may have an influence on the reactions of other substituents. For example, microbial transformation rate constants for the oxidation of phenols to catechols were correlated with the van der Waal radius of the molecules.[52] Microbial rate constants of the hydrolysis of chlorinated carboxylic acid esters, having a fixed aromatic moiety and an increasing length of the alkyl component (2,4-D esters), increased with increasing length of alkyl substituents and with increasing n-octanol/water partition coefficients of the chemicals.[53]

C. Soil-Bound Residues

Binding of pesticides to natural soil constituents is of paramount importance for their ecotoxicological evaluation. Binding is accompanied by a drastic change in pesticide mobility, leaching, and bioavailability including biodegradation and uptake by plants and soil fauna. Multiple binding mechanisms are effective between pesticides and soil constituents (namely, physical, physicochemical, and chemical), most of them being widely unexplored. Pesticides may be bound

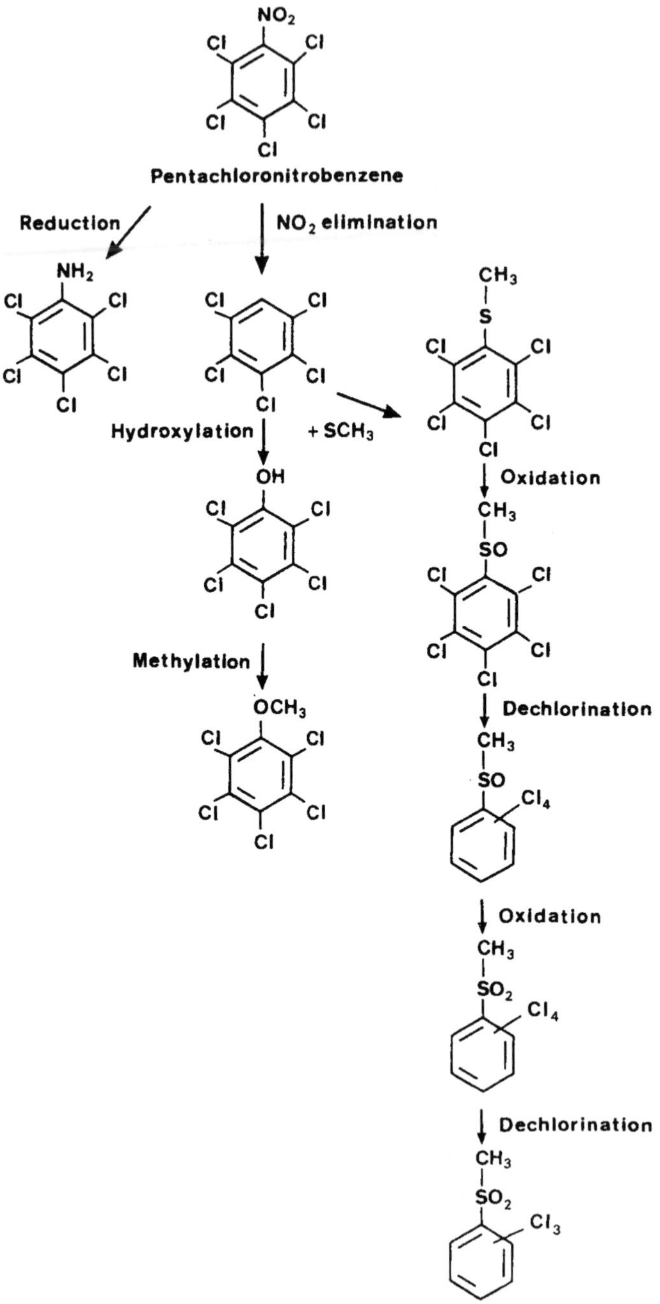

Figure 1. Conversion pathways of pentachloronitrobenzene (quintozene) in soil. [From Scheunert, I. in *Chemistry of Plant Protection, Vol. 8,* W. Ebing, Ed. (Berlin/Heidelberg: Springer, 1992), p. 23.]

Table 4. Proposed Binding Mechanisms of Pesticides to Natural Soil Constituents

Binding fraction	Proposed binding mechanism	Examples	Methods for liberation and/or characterization
Liquid phase	Various	Terbutylazine, pendimethalin	Desorption, hydrolysis
Inorganic solid phase	Deposition into clay minerals	Diquat, paraquat	Conc. H_2SO_4
Organic solid phase	Deposition in cavities between organic macromolecules	Atrazine, prometryn, δ-methrin, diuron, 2,4-D, methyl-parathion, dieldrin, carbofuran	High temperature distillation, supercritical fluid extraction
	Covalent binding to organic macromolecules	Chlorinated anilines, chlorinated phenols	Hydrolysis, degradative reactions; model syntheses

Source: Huber et al.[10] and Scheunert.[19]

in the liquid as well as in the solid soil fraction. In the first case, binding is to low molecular weight, water-soluble humic substances; mobility in soil and leaching from soil are promoted, resulting in the presence of pesticides and their conversion products in leachate, even if the nonbound pesticide is water-insoluble.

In contrast, in the second case mobility in soil and leaching from soil are strongly reduced; the same applies to degradability and bioavailability. Even the extractability of pesticide residues by organic solvents is strongly inhibited; therefore, these residues are called "unextractable" soil-bound residues. Due to the steady molecular changes and rearrangements within humic macromolecules, there is a steady transition between the residues bound to lower and to higher molecular weight humic substances.[54] Table 4 presents some proposed binding mechanisms in the liquid and solid soil fraction.

The binding of pesticides to insoluble, high molecular weight humic material is by multiple mechanisms that are poorly understood. In many cases, the xenobiotic may be entrapped in cavities of the organic macromolecule and is released when the humic acids are dissolved in dilute alkali, e.g., in the case of the aldrin metabolite dihydrochlordene dicarboxylic acid. Other methods to dissolve unextractable xenobiotic residues from soil are high temperature distillation and supercritical fluid extraction. It is assumed that the bound residues released by these two methods discussed are not fixed by covalent bonds. However, between xenobiotics and soil organic matter covalent bonds may be formed, from which the xenobiotic can be liberated in the laboratory only by chemical methods (such as hydrolysis or degradative oxidation) or which are not accessible even to hydrolytic or oxidative attack. Potential types of binding by abiotic and biotic reactions have been discussed in the previous paragraphs.

When considering the binding of organic chemicals in soil, it should not be overlooked that inorganic soil constituents also may play a significant role in binding. For more details of binding of pesticides to soil fractions, see Chapter 13.[55]

The significance of unextractable pesticide residues in soil has been discussed controversially. It is generally adopted that their significance depends on potential release of small amounts of soluble pesticide-derived material which might be available to plant uptake, to leaching, to groundwater pollution, and to effects on microflora of the soil and thus on soil fertility. Solubilization of unextractable residues may be performed by soil microorganisms (e.g., prometryn, parathion, chlorinated anilines, and phenols), by earthworms, or by UV irradiation. For this release, various mechanisms are discussed.[19]

Nonextractable residues that are not covalently bound probably are released in the same chemical form as that which was initially bound. This may be the pesticide itself or a metabolite formed in soil before binding was performed.

For the partial degradation of polymers containing covalently bound pesticide residues, three possibilities are under discussion. First, if within the polymer that bond is cleaved by which the residue had been attached, the original residue is released, as is the case with noncovalently bound residues.[56,57] Second, cleavage of the polymer to oligomers with xenobiotics attached has been reported, e.g., of humic-3,4-dichloroaniline complexes by the fungus *Penicillium frequentans* which is able to break up large humic molecules into fragments.[57] It is uncertain whether these oligomers are similar to those dimeric or trimeric building blocks from which the polymers probably had been formed. Third, degradation of the polymer may occur by biochemical reactions different from those responsible for their formation. Thus, the polymer is cleaved at a site other than the original bond between the pesticide and the natural molecule. This has been demonstrated by the degradation of synthetic model compounds containing β-arylether structures.[58] These structures represent the most abundant bond type in the peroxidase-catalyzed polymerization of coniferylalcohol.[59]

Finally, the xenobiotic moiety of the soil-bound residue may be mineralized to CO_2. This product, of course, must not be regarded as a xenobiotic, and its further fate in the terrestrial ecosystem is without importance from the ecotoxicological point of view.[19]

A rough predictive estimation of the level of soil-bound pesticide residues is possible by their chemical structure characteristics. Whereas the formation rate decreases with increasing chlorine content of a pesticide,[60] a high formation rate may be expected if aromatic hydroxyl or amino groups are present in the molecule which then is similar to a humic acid monomer and may be incorporated into soil organic matter.

IV. TRANSPORT OF TRANSFORMATION PRODUCTS

Since upon biotic or abiotic transformation the alteration in substituents implies an alteration in pesticides physical properties and thus in their mobility in soil, enhanced attention has to be paid to potential transformation products transported

Table 5. Conversion Products of Pesticides,
Transported from Soil to Other Media

Receiving medium	Pesticide	Conversion products	Ref.
Air	Parathion	Oxy analogue	18
Fog water	Parathion	Oxy analogue	18
	Methidathion	Oxy analogue	18
	Chlorpyrifos	Oxy analogue	18
Drainage water	Atrazine	Deethyl analogue	61
	Terbutylazine	Deethyl analogue	61
Groundwater	Atrazine	Deethyl analogue	61
	Simazine	Deethyl analogue	61
	Aldicarb	Sulfone	62

from soil to air or water. So far, information on the presence of pesticide conversion products in leachate or in the atmosphere is limited. A few examples are listed in Table 5.

The majority of transformation products of pesticides are more hydrophilic than the parent compounds and thus more accessible to leaching. This applies even more to binding products between pesticides or their metabolites to dissolved organic carbon in the liquid soil fraction. When [^{14}C]terbutylazine was applied to lysimeters under outdoor conditions, ^{14}C in the percolation water at a 1-m depth contained only a very small proportion of the free parent pesticide (1.9%). The known conversion products — deethyl-terbutylazine and de-t-isobutyl-ter-butylazine — also were only a small fraction of total ^{14}C in the leachate (3.8 or 0.3%, respectively), at least in a free, nonbound state. The remaining ^{14}C was unknown transformation products, most of which were fully water-soluble and had a high molecular weight, i.e., they were adducts to dissolved organic carbon.[10,63] Hardly any attention has been paid so far to this phenomenon of transport of pesticide residues into groundwater.

V. MINERALIZATION

Mineralization of pesticides means their complete degradation to small inorganic molecules such as CO_2, CO, H_2O, NH_3, H_2S, Cl^-, and other species. It is the only way to eliminate a xenobiotic compound from the environment. A final product resulting from cometabolic activity of certain species may be degraded further by other species groups, or may be accessible to abiotic attack. Thus, mineralization in soil is a complex process in which various biotic and abiotic factors are involved.

The quantification of total degradation of foreign compounds in soil is very important for their ecotoxicological evaluation. The best method to perform it is the quantitative determination of $^{14}CO_2$ evolved from ^{14}C-labeled substances in the laboratory. The time course of biomineralization is a major topic of current research into biodegradation of man-made organic chemicals. Information on these kinetics is indispensable for an examination of the extent and relevance of mineralization.

In biomineralization experiments measuring $^{14}CO_2$, many time curves show apparent lag phases at the beginning followed by an exponential second phase and a third phase where mineralization decreases continuously until a complete stop is reached; complete mineralization is not achieved. This sigmoid curve course of $^{14}CO_2$ release from ^{14}C-labeled organic compounds is reported frequently in the literature.[64-66] It is a characteristic of biological systems involving growth and/or adaptation in soil. Scow et al.[65] showed that the lag phase accounts for an adaptation of soil microflora to the foreign compound. The third phase may represent the stage where a substrate level is reached which is too low to sustain the metabolizing microbial fraction. It might also represent the mineralization of formerly assimilated carbon from the xenobiotic either by normal respiration, or by decay after the death of the microorganisms. This evolution of carbon dioxide continues also after degradation of the original xenobiotic has discontinued.

If relative biomineralization rates are required in order to compare different chemicals with each other or to compare a new chemical with a well-known reference compound, biomineralization studies are conducted under standardized conditions in the laboratory; measurements of $^{14}CO_2$ then are restricted to a few time intervals. Table 6 gives some data obtained after shaking ^{14}C-labeled pesticides in a soil-water suspension under aerobic or anaerobic conditions at 35°C.[67] $^{14}CO_2$ was trapped by drawing air once a day through the apparatus and absorption of $^{14}CO_2$ in a scintillation liquid containing an organic base. The table lists the sum of $^{14}CO_2$ after 5 days and at the end of the experiment after 14—56 days.

The table reveals that all these compounds are degraded, although most of them only at a very low rate. The differences in degradation rates between aerobic and anaerobic conditions are low. For some chlorinated compounds, degradation is somewhat better under anaerobic than under aerobic conditions. The reason is the first step of degradation, namely, reductive dechlorination, which is favored under anaerobic conditions.

Table 7 presents mineralization rates in another laboratory system — a closed soil system containing barley.[68-73] In general, $^{14}CO_2$ formation is low for highly chlorinated compounds (as shown in the lower part of the table) and higher for lower chlorinated compounds. However, exceptions are evident due to the influence of other substituents in the molecule. Thus, the mineralization of pentachlorophenol is rapid although it has five chlorines in the molecule. The phenolic hydroxy group might be the reason for this rapid degradability. The herbicide trichloroacetic acid also may be regarded as readily biodegradable, whereas the other pesticides listed here in normal soil-plant systems are rather persistent.

Table 6. Biodegradation of ^{14}C-Labeled Pesticides in a Soil-Water Suspension Under Aerobic and Anaerobic Conditions (35°C)

^{14}C-Labeled Chemical applied	Experimental time (days)	Conditions	$^{14}CO_2$ after 5 days (% of ^{14}C applied)	$^{14}CO_2$ at the end of experiment (% of ^{14}C applied)	^{14}C in water at the end of experiment (% of ^{14}C applied)
4-Chloroaniline	56	Aerobic	1.5	3.0	14.6
		Anaerobic	n.m.	2.3	22.1
Lindane	42	Aerobic	0.4	1.9	3.1
		Anaerobic	0.4	3.0	2.2
DDT	42	Aerobic	0.1	0.8	2.4
		Anaerobic	0.3	0.7	2.0
2,4-D	14	Aerobic	0.1	0.5	89.1
		Anaerobic	0.2	0.7	83.4
2,6-Dichlorobenzonitrile	42	Aerobic	<0.01	0.5	38.9
		Anaerobic	<0.01	<0.01	36.8
Hexachlorobenzene	14	Aerobic	<0.01	0.4	1.9
		Anaerobic	<0.01	0.2	2.1

Note: n.m. = not measured; 2,4-D = 2,4-dichlorophenoxyacetic acid.

Table 7. Biodegradation of ^{14}C-Labeled
Pesticides in Soil-Plant Systems

^{14}C-Labeled pesticide applied	% ^{14}CO$_2$
Pentachlorophenol	13.9
Trichloroacetic acid	7.7
Aldrin	1.6
4-Chloroaniline	1.1
Chlortoluron	0.68
Atrazin	0.48
Kelevan	0.07
Trifluralin	0.05
Kepone	0.01
Dieldrin	0.01
p,p'-DDT	0.01
Terbutylazine	<0.01
Hexachlorobenzene	<0.01

Note: At room temperature for 7 days.

VI. CONCLUSION

It may be concluded that, despite a great number of data available, the scientific community today is far from a quantitative understanding of all processes affecting pesticide persistence in soil. Much more information still has to be obtained both for individual processes and for their interactions and interdependencies. However, the information today in some cases gives the possibility of predicting some of the factors affecting the fate of pesticides in soil.

REFERENCES

1. Scheunert, I. "Physical and Physico-Chemical Processes Governing the Residue Behaviour of Pesticides in Terrestrial Ecosystems," in W. Ebing, Ed. *Chemistry of Plant Protection, Vol. 8* (Berlin/Heidelberg: Springer, 1992), p. 1.
2. Adams, R.S. Factors Influencing Soil Adsorption and Bioactivity of Pesticides, *Res. Rev.* 47:1 (1973).
3. Mingelgrin, U.and Z. Gerstl. Reevaluation of Partitioning as a Mechanism of Nonionic Chemicals Adsorption in Soils, *J. Environ. Qual.* 12:1 (1983).
4. Jury, W.A., A.M. Winer, W.F. Spencer, and D.D. Focht. "Transport and Transformations of Organic Chemicals in the Soil-Air-Water Ecosystem," in *Reviews of Environmental Contamination and Toxicology, Vol. 99* (Berlin: Springer, 1987), p. 119.
5. Bailey, G.W. and J.L. White. Factors Influencing the Adsorption, Desorption and Movement of Pesticides in Soil, *Res. Rev.* 32:29 (1970).

6. Scheunert, I. and W. Klein. "Predicting the Movement of Chemicals Between Environmental Compartments (Air-Water-Soil-Biota)," in *Appraisal of Tests to Predict the Environmental Behaviour of Chemicals, SCOPE 25*, P. Sheehan, F. Korte, W. Klein, and Ph. Bourdeau, Eds. (Chichester: John Wiley & Sons, 1985), p. 285.

7. Weber, W.J., Jr. *Physicochemical Processes for Water Quality Control* (New York: Wiley Interscience, 1972).

8. Von Oepen, B., W. Kördel, and W. Klein. "Models for Estimating Adsorption to Soils — Scope and Limitations," in *Book of Abstracts, Seventh International Congress of Pesticide Chemistry, Vol. 3*, H. Frehse, E. Kesseler-Schmitz, and S. Conway, Eds. (Hamburg: IUPAC-GDCH, 1990), p. 145.

9. Klein, M., W. Kördel, W. Klein, and A.W. Klein. "Sensitivity of Soil Leaching Simulation Models for Environmental Variables," in *Book of Abstracts, Seventh International Congress of Pesticide Chemistry, Vol. 3*, H. Frehse, E. Kesseler-Schmitz, and S. Conway, Eds. (Hamburg: IUPAC-GDCH, 1990), p.421.

10. Huber, S., I. Scheunert, U. Dörfler, and F.H. Frimmel. Zum Einfluß des gelösten organischen Kohlenstoffs (DOC) auf das Mobilitätsverhalten einiger Pestizide, *Acta Hydrochim. Hydrobiol.* 20:74 (1992).

11. Cohen, S.Z., S.M. Creeger, R.F. Carsel, and C.G. Enfield. "Potential Pesticide Contamination of Groundwater from Agricultural Use," in *Treatment and Disposal of Pesticide Wastes*, ACS Symposium Series 259, R.F. Krüger and J.N. Seiber, Eds. (Washington, DC: American Chemical Society, 1984), p. 297.

12. "Preliminary Assessment of Suspected Carcinogens in Drinking Water," U.S. Environmental Protection Agency, Washington, DC (1975).

13. Spencer, W.F., W.J. Farmer, and M.M. Cliath. Pesticide Volatilization, *Res. Rev.* 49:1 (1973).

14. Hamaker, J.W. "Diffusion and Volatilization," in *Organic Chemicals in the Soil Environment, Vol. 2*, C.A.I. Goring and J.W. Hamaker, Eds. (New York: Marcel Dekker, 1972), p. 341.

15. Spencer, W.F., M.M. Cliath, W.A. Jury, and L.Z. Zhang. Volatilization of Organic Chemicals from Soil as Related to their Henry's Law Constants, *J. Environ. Qual.* 17:504 (1988).

16. Herterich, R. Atrazin — Atmosphärischer Eintrag und Immissions-Konzentrationen, *UWSF-Z. Umweltchem. Ökotox.* 3:196 (1991).

17. Oberwalder, Ch., H. Giessl, L. Irion, J. Kirchhoff, and K. Hurle. Pflanzenschutzmittel im Niederschlagswasser, *Nachrichtenbl. Dtsch. Pflanzenschutzd.* 43:185 (1991).

18. Glotfelty, D.E., J.N. Seiber, and L.A. Liljedahl. Pesticides in Fog, *Nature* 325:602 (1987).

19. Scheunert, I. "Transformation and Degradation of Pesticides in Soil, in *Chemistry of Plant Protection, Vol. 8*, W. Ebing, Ed. (Berlin/Heidelberg: Springer, 1992), p. 23.

20. Guth, J.A. "Experimental Approaches to Studying the Fate of Pesticides in Soil," in *Progress in Pesticide Biochemistry*, D.H. Hutson and T.R. Roberts, Eds. (Chichester: John Wiley & Sons, 1981), p. 85.

21. Wolfe, N.L., U. Mingelgrin, and G.C. Miller. "Abiotic Transformations in Water, Sediments, and Soil," in *Pesticides in the Soil Environment: Processes, Impacts, and Modeling*, H.H. Cheng, Ed. (Madison, WI: Soil Science Society of America, 1990), p. 103.

22. Wang, T.S.C., S.W. Li, and Y.L. Ferng. Catalytic Polymerization of Phenolic Compounds by Clay Minerals, *Soil Sci.* 126:15 (1978).

23. Adrian, P., E.S. Lahaniatis, F. Andreux, M. Mansour, I. Scheunert, and F. Korte. Reaction of the Soil Pollutant 4-Chloroaniline with the Humic Acid Monomer Catechol, *Chemosphere* 18:1599 (1989).

24. Turner, N.J. and M.E. Corden. Decomposition of Sodium N-Methyldithiocarbamate in Soil, *Phytopathology* 53:1388 (1963).

25. Miller, P.M. and R.J. Lukens. Deactivation of Sodium N-Methyldithiocarbamate in Soil by Nematocides Containing Halogenated Hydrocarbons, *Phytopathology* 56:967 (1966).

26. Brown, C.R. and S.F. Rahmann. *Soil Sci. Soc. Am. J.* 41:141 (1977).

27. Nash, R.G. and W.F. Harris. *Weed Sci. Soc. Am.* (Abstr.) 240 (1968).

28. Mansour, M., I. Scheunert, and F. Korte. "Fate of Persistent Organic Compounds in Soil and Water," in *Migration and Fate of Pollutants in Soils and Subsoils,* NATO ASI Series 6, Ecological Sciences Vol. 32, D. Petrozzelli, F.G. Helfferich, Eds. (Berlin/Heidelberg: Springer Verlag, 1993), p. 111–132.

29. G.T. Brooks. "Pathways of Enzymatic Degradation of Pesticides," in *Environmental Quality and Safety, Vol. 1,* F. Coulston and F. Korte, Eds. (Stuttgart: Georg Thieme Verlag/Academic Press, 1972), p. 106.

30. Kearney, P.C., J.K. Plimmer, and F.B. Guardia. Mixed Chloroazobenzene Formation in Soil, *J. Agric. Food Chem.* 17:1418 (1969).

31. Bartha, R. Pesticide Interaction Creates Hybrid Residue, *Science* 166:1299 (1969).

32. Liu, S.-Y., R.D. Minard, and J.-M. Bollag. Coupling Reactions of 2,4-Dichlorophenol with Various Anilines, *J. Agric. Food Chem.* 29:253 (1981).

33. Reineke, W. Der Abbau von chlorierten Aromaten durch Bakterien: Biochemie, Stammentwicklung und Einsatz zur Boden- und Abwasserbehandlung, *Forum Mikrobiol.* 12:402 (1989).

34. Golovleva, L.A. and G.K. Skryabin. "Microbial Degradation of DDT," in *Microbial Degradation of Xenobiotics and Recalcitrant Compounds,* T. Leisinger, A.M. Cook, R. Hütter, J. Nüesch, Eds. (London: Academic Press, 1981), p. 287.

35. Maule, A., S. Plyte, and A.V. Quirk. Dehalogenation of Organochlorine Insecticides by Mixed Anaerobic Microbial Populations, *Pestic. Biochem. Physiol.* 27:229 (1987).

36. Murthy, N.B.K., D.D. Kaufman, and G.F. Fries. Degradation of Pentachlorophenol (PCP) in Aerobic and Anaerobic Soil, *J. Environ. Sci. Health, Part B* 14:1 (1979).

37. Weiss, U.M., I. Scheunert, W. Klein, and F. Korte. Fate of Pentachlorophenol-^{14}C in Soil under Controlled Conditions, *J. Agric. Food Chem.* 30:1191 (1982).

38. Jagnow, G., K. Haider, and P.C. Ellwardt. Anaerobic Dechlorination and Degradation of Hexachlorocyclohexane Isomers by Anaerobic and Facultative Anaerobic Bacteria, *Arch. Microbiol.* 115:285 (1977).

39. Allan, J. Loss of Biological Efficiency of Cattle-Dipping Wash Containing Benzene Hexachloride, *Nature* 175:1132 (1955).

40. Fries, G.F. "Degradation of Chlorinated Hydrocarbons Under Anaerobic Conditions," in *Fate of Organic Pesticides in the Aquatic Environment,* Advan. Chem. Ser. III (Washington, DC: American Chemical Society, 1972), p. 256.

41. Motosugi, K. and K. Soda. Microbial Degradation of Synthetic Organochlorine Compounds, *Experientia* 39:1214 (1983).

42. Belser, N.O. and C.E. Castro. Biodehalogenation — The Metabolism of the Nematocides cis- and trans-3-Chloroallyl Alochol by a Bacterium Isolated from Soil, *J. Agric. Food Chem.* 19:23 (1971).
43. Rott, B., S. Nitz, and F. Korte. On the Microbial Decomposition of Sodium Pentachlorophenolate, *J. Agric. Food Chem.* 27:306 (1979).
44. Apajalahti, J.H.A. and M.S. Salkinoja-Salonen. Dechlorination and para-Hydroxyltion of Polychlorinated Phenols by Rhodococcus chlorophenolicus, *J. Bacteriol.* 169:675 (1987).
45. Bekhi, R.M. and S.U. Khan. Degradation of Atrazine by Pseudomonas: N-Dealkylation and Dehalogenation of Atrazine and its Metabolites, *J. Agric. Food Chem.* 34:746 (1986).
46. Guenzi, W.D. and W.E. Beard. DDT Degradation in Flooded Soil as Related to Temperature, *J. Environ. Qual.* 5:391 (1976).
47. Kohli, J., I. Weisgerber, W. Klein, and F. Korte. Contributions to Ecological Chemistry CVII. Fate of Lindane-^{14}C in Lettuce, Endives and Soil under Outdoor Conditions, *J. Environ. Sci. Health,* Part B 11:23 (1976).
48. Khalil, A.M. "Aufnahme und Metabolismus von Hexachlorbenzol-^{14}C und Pentachlornitrobenzol-^{14}C in Regenwürmern," Doctoral Thesis, Technical University of Munich, Germany (1990).
49. Müller, R. and F. Lingens. Mikrobieller Abbau Halogenierter Kohlenwasserstoffe: Ein Beitrag zur Lösung vieler Umweltprobleme? *Angew. Chem.* 98:778 (1986).
50. Karlson, P. *Kurzes Lehrbuch der Biochemie,* 7th ed. (Stuttgart: Thieme, 1970).
51. Cairns, T., E.G. Siegmund, and F. Krick. Identification of Several new Metabolites from Pentachloronitrobenzene by Gas Chromatography/Mass Spectrometry, *J. Agric. Food Chem.* 35:433 (1987).
52. Paris, D.F., N.L. Wolfe, W.C. Steen, and G.L. Baughman. Effect of Phenol Molecular Structure on Bacterial Transformation Rate Constants in Pond and River Samples, *Appl. Environ. Microbiol.* 45:1153 (1983).
53. Paris, D.F., N.L. Wolfe, W.C. Steen. Microbial Transformation of Esters of Chlorinated Carboxylic Acids, *Appl. Microbiol.* 47:7 (1984).
54. Scheunert, I., M. Mansour, and F. Andreux. Binding of Organic Pollutants to Soil Organic Matter, *Int. J. Environ. Anal. Chem.* 46:189 (1992).
55. Andreux, F., I. Scheunert, P. Adrian, and M. Schiavon. "The Binding of Pesticide Residues to Natural Organic Matter, Their Movement and Their Bioavailability," in *Fate and Prediction of Environmental Chemicals in Soils, Plants, and Aquatic Systems* Ann Arbor, MI: Lewis Publishers, 1993).
56. Dec, J. and J.-M. Bollag. Microbial Release and Degradation of Catechol and Chlorophenols Bound to Synthetic Humic Acid, *Soil Sci. Soc. Am. J.* 52:1366, (1988).
57. Hsu, T.-S. and R. Bartha. Biodegradation of Chloroaniline-Humus Complexes in Soil and in Culture Solution, *Soil Sci.* 118:213, (1974).
58. Engelhardt, G., P.R. Wallnöfer, and H.-G. Rast. "Bacterial Degradation of Veratrylglycerol-β-arylethers as Model Compounds for Soil-bound Pesticide Residues," in *Microbial Degradation of Xenobiotics and Recalcitrant Compounds,* T. Leisinger, A.M. Cook, R. Hütter, J. Nüesch, Eds. (London: Academic Press, 1981), p. 293.
59. Rast, H.G., G. Engelhardt, W. Ziegler, and P.R. Wallnöfer. *FEMS Microbiol. Lett.* 8:259 (1980).

60. Klein, W. and I. Scheunert. "Bound Pesticide Residues in Soil, Plants and Food with Particular Emphasis on the Application of Nuclear Techniques," in *Agrochemicals: Fate in Food and the Environment*, IAEA-SM-263/39 (Vienna: International Atomic Energy Agency, 1982), p. 177.

61. Hurle, K., H. Giessl, and J. Kirchhoff. "Über das Vorkommen einiger ausgewählter Pflanzenschutzmittel im Grundwasser," in *Grundwasserbeeinflussung durch Pflanzenschutzmittel, Schriftenreihe des Vereins für Wasser-, Boden- und Lufthygiene 68*, G. Milde and P. Friesel, Eds. (Stuttgart: Gustav Fischer, 1987), p. 169.

62. Ahlsdorf, B., C. Ehrig, E. Zeeb, U. Müller-Wegener, and G. Milde. "Grundwasserbelastung durch Pflanzenschutzmittel — Felduntersuchungen in oberflächennahen und tieferen Grundwässern," paper presented at the Jahrestagung der Deutschen Bodenkundlichen Gesellschaft, Münster, Germany, September 2–10, 1989.

63. Dörfler, U. and I. Scheunert. BMFT-Verbundvorhaben 02-WT 89137 "Aufklärung der für den Pflanzenschutzmitteleintrag verantwortlichen Vorgänge, insbesondere im Hinblick auf die Trinkwasserversorgung," 3. Zwischenbericht, German Federal Ministry of Research and Technology (1992).

64. Simkins, S. and M. Alexander. Models for Mineralization Kinetics with the Variables of Substrate Concentration and Population Density, *Appl. Environ. Microbiol.* 47:1299 (1984).

65. Scow, K.M., S. Simkins, and M. Alexander. Kinetics of Mineralization of Organic Compounds at low Concentrations in Soil, *Appl. Environ. Microbiol.* 5:1028 (1986).

66. Brunner, W. and D.D. Focht. Three-half-order Kinetic Model for Microbial Degradation of Added Carbon Substrates in Soil, *Appl. Environ. Microbiol.* 47:167 (1984).

67. Scheunert, I., D. Vockel, J. Schmitzer, and F. Korte. Biomineralization Rates of ^{14}C-labelled Organic Chemicals in Aerobic and Anaerobic Suspended Soil, *Chemosphere* 16:1031 (1987).

68. Topp, E.M. "Aufnahme von Umweltchemikalien in die Pflanze in Abhängigkeit von physikalisch-chemischen Stoffeigenschaften," Dissertation, Technische Universität München, Germany (1986).

69. Schroll, R.E. "Aufnahme ^{14}C-markierter Chemikalien und Radionuklide aus Boden in Pflanzen in Abhängigkeit von Pflanzeneigenschaften," Dissertation, Technische Universität München, Germany (1989).

70. Kloskowski, R. "Konzeption und Optimierung eines Pflanze/Boden Testsystems zur Bewertung von Umweltchemikalien," Dissertation, Technische Universität München, Germany (1981).

71. Cao, G. Unpublished results (1991).

72. Zhang, Q. Unpublished results (1991).

73. Schroll, R., T. Langenbach, G. Cao, U. Dörfler, P. Schneider, and I. Scheunert. Fate of ^{14}C-Terbutylazine in Soil-Plant-Systems, *Sci. Total Environ.* 123/124:377 (1992).

CHAPTER 2

Prediction of Uptake of Some Aromatics and Pesticides by Soil

Gyula Szabó, S. Lesley Prosser, and Robert A. Bulman

ABSTRACT

Soil adsorption coefficients, K_{OC}, for a series of aromatics and pesticides have been determined from high-performance liquid chromatography (HPLC) capacity factors, k', by using a new HPLC immobilized phase which bears immobilized humic acid. From the values for K_{OC} it appears this new HPLC phase is superior to other procedures — partition into n-octanol and retention on an HPLC ethyl silica phase — for modeling the sorption of organic chemicals on soil or sediment.

I. INTRODUCTION

The migration of chemicals through soil and the tendency for the chemical to accumulate in the food chain have been shown to be related to the "basic hydrophobicity" of the molecule.[1] Several correlations have been established between the following standardized partitioning parameters: n-octanol: water ratio (K_{ow}), the soil sorption coefficient based on organic carbon (K_{oc}), the bioconcentration factor (BCF), and the water solubility (S).[2-7]

Although direct methods of measuring K_{oc} are preferred, its estimation from other physical properties such as water solubility (S) and the n-octanol/water

0-87371-616-7/93/$0.00 + $.50

Table 1. Reference Chemicals with Reliable log K_{oc} and
log K_{ow} Values Obtained from the Literature

Chemicals	Log K_{oc}	Ref.	Log K_{ow}	Ref.
Benzene	1.91	3	2.11	3
Toluene	2.18	16	2.65	17
o-Xylene	2.34[a]	11	3.13	17
Ethylbenzene	2.41[a]	11	3.13	17
1,2-Dichlorobenzene	2.54	2	3.39	2
Propylbenzene	2.86[a]	11	3.69	17
Naphthalene	3.11	3	3.36	3
Lindane (τ BHC)	3.30	2	3.80	2
Butylbenzene	3.40[a]	11	4.28	17
Phenanthrene	4.28[a]	11	4.57	3
Anthracene	4.41	3	4.54	3
Pyrene	4.83	2	5.18	17
Methoxychlor	4.90	2	5.08	2
pp' DDT	5.38	2	6.19	2

[a] Converted from literature log K_p value using the relationship
$K_{oc} = K_p \times 100/\%$ organic carbon.

partition coefficient (K_{ow}) can also be used. However, many of these correlations were developed for specific classes of compounds and are not usually applicable to other classes of compounds. In addition for many compounds, K_{ow} or S values are unavailable because experimental determination of these parameters are difficult, particularly for highly hydrophobic compounds. As values for K_{ow} and K_{oc} are difficult to determine, alternative methods for their determination have been considered.

Prominent among these alternative methods has been their determination by reverse-phase high-performance liquid chromatography (RP-HPLC).[8-10] By using RP-HPLC with a cyanopropyl column and a mobile phase of high water content, it has been possible to measure sediment/water adsorption coefficients for alkyl benzenes, polyaromatic hydrocarbons, and some pesticides.[11,12] These investigations demonstrated that the relationship between HPLC capacity factors (k') and K_p or K_{oc} could be satisfactorily described by a simple correlation equation, whereas the octadecyl (ODS) column required the use of three correlation equations to describe the relationship between these properties. In an extension of these investigations, Szabó et al.[13] have shown that an ethyl-silica phase also affords prediction of the soil/water partition coefficient. It is easily seen that there is some uncertainty for determination of K_{oc} or K_p using different types of chromatographic materials, as these materials have polarities which might not reflect the polarity of the organic material in soil.

So that we might model the adsorptive surface properties of the soil we have prepared an HPLC packing material bearing immobilized humic acid and studied the effect of changing the mobile phase water content on the correlation between log k' and log K_{oc}. To evaluate this new phase we have constructed a calibration curve using log K_{oc} values, listed in Table 1, and experimentally determined values for the theoretical HPLC capacity factor, k'_w. From this calibration curve we have redetermined log K_{oc} values for those chemicals listed in Table 1; in addition, we have calculated log K_{oc} values for other chemicals. By way of a comparison we have also determined log K_{oc} on an ethyl-silica phase we have synthesized.

II. EXPERIMENT

A. Apparatus and Materials

Chromatographic retention data were measured with an LKB 2150 solvent delivery system and LKB Wavescan diode-array detector. Sample introduction was via a Rheodyne 7125 injection valve fitted with 20 μL loop. Chromatographic data were collected using commercially available software (Wavescan, LKB) and recorded on an Olivetti M24 personal computer. The ethyl-phase column (250 × 4.6 mm) was packed by Bio-Separation Technologies Co. (Budapest, Hungary). The immobilized humic acid packing material was packed in 250 × 4.6 mm stainless steel tube by Jones Chromatography (Hengoed, Wales). Hypersil WP 300 5 μm was obtained from Shandon, (Cheshire); 3-aminopropyltriethoxysilane (Fluka Chemicals Ltd.) and triethylchlorosilane (Petrach System) were used as received. Humic acid was purchased from Aldrich Chemical Co. All other chemicals obtained from commercial sources were used as received.

B. Chromatography

Mobile phases, HPLC grade methanol and water, were mixed volume per volume and nitrogen displaced by helium. The test solutes were dissolved in methanol at a concentration of 0.1 mg/mL by weight. Flow rates were 0.8 mL/min as usual. A laboratory temperature of 20–23°C was used for all HPLC measurements. The mobile phase methanol content was changed from 70 to 50% by 5% steps. Methanol was used for the measurement of the retention time (t_0) of an unretained compound. The relationship:

$$k' = t_r - t_0/t_0$$

was used to calculate the capacity factor, k', from the retention time (t_r) of each compound. All capacity factors reported are the mean of at least three measurements.

C. Preparation of Immobilized Humic Acid Silica Gel

Dried silica gel (10 g) was refluxed with 5% 3-aminopropyltriethoxysilane in anhydrous toluene.[14] The resulting aminopropyl silica gel was removed by filtration; washed with toluene, methanol, and water; washed again with methanol;

and dried. Activation of the derivatized silica was achieved by stirring the gel in 5% aqueous glutaraldehyde (10 volumes) for 5 hr to produce a gel which on isolation was washed with 15 volumes of distilled water. This purified gel was reacted with 100 mL of 1% aqueous solution of humic acid, pH 7.5, for 8 hr at ambient temperature. After the immobilized humic acid gel had been washed with 10 volumes of 0.5 M phosphate buffer and distilled water it was treated with 0.1 M buffered ethanolamine, pH 7.5, for 3 hr. The reaction product was washed with a large excess of distilled water and dried to yield a dark-brown product. Elemental analysis of this new stationary phase was performed in the Micro Analytical Laboratory at the University of Manchester.

D. Preparation of Ethyl Silica Phase

The preparation of the ethyl packing material was based on Unger's procedure.[15] Basically, moisture-free dried silica gel, 10 g, was refluxed for 8 hr with 5% triethylchlorosilane in anhydrous toluene; and the isolated ethyl silica gel was washed with toluene, methanol, and water and then again with methanol. This packing material was packed as a slurry under high pressure in a stainless steel column (250 × 4.6 mm) by Bio-Separation Technologies (Budapest).

E. Selection of Log K_{oc} and Log K_{ow} Values

The literature contains only a limited number of compounds with reliable log K_{oc} values. In Table 1 we have compiled log K_{oc} and log K_{ow} values for the compounds used in this study. Compilation of this table was achieved using the well-established criteria, originally used by Brooke et al.[16] for selection of reliable log K_{ow} values.

All the reported capacity factors are the mean of at least three measurements. The correlation analysis for all compounds was made by linear regression analysis of log k' or log K_{ow} vs log K_{oc} and a least-squares fit routine was used for curve fitting.

Elemental analysis of the humic acid phase revealed a C, H, and N content of 4.7, 0.7, and 0.5%, respectively. Figure 1 shows the correlation between log K_{oc} and log K_{ow} values using the data available from Table 1 and giving the relationship:

$$\log K_{oc} = 1.002 \log K_{ow} - 0.526 \quad r^2 = 0.937 \quad n = 14 \quad (1)$$

The slope parameter (1.002) compares well with values reported elsewhere by Means et al. and Karickoff[1,2] who reported a value of 1.0 for various substrates and solutes.

In order to eliminate selective solute-solvent interactions,[18] we have used log k'_w, the capacity factor obtained by extrapolation of retention data from binary

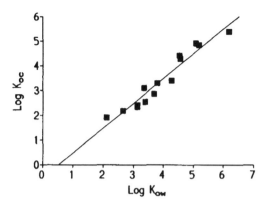

Figure 1. Relationship between soil/water partition coefficients (K_{oc}) and n-octanol/water partition coefficients (K_{ow}).

eluents to 100% water, instead of using log k', the capacity factor obtained from binary eluents. Snyder et al.[19] showed that linear equation:[2]

$$\log k' = \log k'_w + S\Phi \qquad (2)$$

can be used to describe the relationship between log k' and log k'_w,

where Φ = volume fraction of organic solvent in the water-organic solvent mixture

k'_w = capacity factor of a solute with pure water as a mobile phase

S = constant for a given solute-eluent combination

The log k'_w of 14 organic compounds calculated from Equation 2 on the immobilized humic acid and ethyl phases are presented in Table 2, along with values of log K_{oc} for these chemicals.

A plot of log K_{oc} vs log k'_w for the chemicals listed in Table 2 yields the calibration curve depicted in Figure 2 for the ethyl-silica stationary phase. Analysis of this calibration curve yields:

$$\log K_{oc} = 1.169 \log k'_w + 1.711 \qquad r^2 = 0.947 \quad n = 14 \qquad (3)$$

A similar treatment of the data for the humic acid phase (Figure 3) yields:

$$\log K_{oc} = 0.938 \log k'_w + 1.756 \qquad r^2 = 0.987 \quad n = 14 \qquad (4)$$

Comparison of Equations 3 and 4 demonstrates that the immobilized humic acid phase gives a better correlation between log K_{oc} and log k'_w than the ethyl

Table 2. Log K_{oc} and Log k'_w Values Obtained from Equation 2, on the Immobilized Humic Acid and Ethyl Phases

Chemical	Log K_{oc}	Log k'_w on humic acid phase	Log k'_w on ethyl phase
Benzene	1.91	0.203	0.159
Toluene	2.18	0.514	0.479
o-Xylene	2.34	0.627	0.759
Ethylbenzene	2.41	0.782	0.739
1,2-Dichlorobenzene	2.54	0.877	0.727
Propylbenzene	2.86	1.107	1.156
Naphthalene	3.11	1.452	0.934
Lindane (τ BHC)	3.30	1.867	1.413
Butylbenzene	3.40	1.407	1.577
Phenanthrene	4.28	2.569	1.859
Anthracene	4.41	2.907	1.961
Pyrene	4.83	3.212	2.374
Methoxychlor	4.90	3.230	2.797
pp' DDT	5.38	4.015	3.491

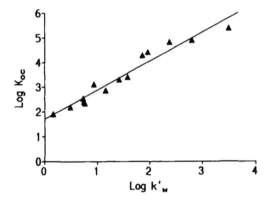

Figure 2. Relationship between soil/water partition coefficients (K_{oc}) and theoretical capacity factors (k'_w) on the ethyl phase.

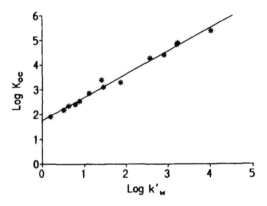

Figure 3. Relationship between soil/water partition coefficients (K_{oc}) and theoretical capacity factors (k'_w) on the immobilized humic acid phase.

Table 3. Log K_{oc} Values Obtained by Using
Immobilized Humic Acid Gel and Differences
from Reported Values

Chemical	Log K_{oc} from HPLC	Difference from reported value
Benzene	1.94	−0.03
Toluene	2.24	−0.06
o-Xylene	2.34	0.00
Pirimicarb (I)	2.37	
Alachlor (H)	2.43	
Ethylbenzene	2.49	−0.08
Carbendazine (F)	2.49	
Benomyl (F)	2.52	
1,4-Dichlorobenzene	2.56	
1,2-Dichlorobenzene	2.58	−0.04
Captan (F)	2.71	
Propylbenzene	2.79	0.07
Butylbenzene	3.10	0.30
Naphthalene	3.12	−0.01
1,2,5-Trichlorobenzene	3.45	
Lindane (I)	3.49	−0.19
Phenanthrene	4.16	0.12
Dieldrin (I)	4.32	
Anthracene	4.48	−0.07
Pyrene	4.77	0.06
Aldrin (I)	4.78	
Methoxychlor (I)	4.79	0.11
DDE	5.40	
pp' DDT (I)	5.53	−0.15

Note: (F) Fungicide; (H) herbicide; (I) insecticide.

phase. On the basis of correlations of K_{oc} vs. K_{ow} and log K_{oc} vs. log k_w', it is more accurate to estimate log K_{oc} from k_w' determined from humic acid phase column than via a single relationship between log K_{oc} and log K_{ow} or log K_{oc} and log k_w' on the ethyl phase. The divergences for the chemicals listed in Table 1 are small for the humic acid phase, suggesting that its adsorptive properties are similar to those of sediment and soil organic matter. Such similarity is predictable because humic acids represent a large proportion of the organic matter in soil. Thus the following order can be generated: organic matter on the soil > immobilized humic acid > ethyl > n-octanol. This is the order we might expect on the basis of the polarity of the media. n-Octanol should be the least polar while sediment or soil organic matter should be the most polar due to the presence of carboxyl, phenolic, and amino moieties in humic substances.

By using Equation 4, log K_{oc} values have been calculated; and these values are listed in Table 3, together with the difference from literature values. In addition we have calculated the values of log K_{oc} for some pesticides and other chemicals which have not been found in the literature.

IV. CONCLUSION

The results of this study establish — provided it is accepted that organic carbon is the most important component for soil adsorption for organic substances — that the log K_{oc} values are determinable from HPLC capacity factors measured on an immobilized humic acid column using reference chemicals for calibration. We suggest the best way to model the soil/water sorption of aromatics and pesticides in the environment is to use log k_w', the capacity factor obtained by extrapolation of retention data from binary eluents to 100% water. For the humic acid phase the correlation between log K_{oc} and log k_w' has been found to be better than that between log K_{oc} and log K_{ow}. Our study shows that the humic acid phase is superior to any commercial HPLC phase for predicting the adsorption coefficients of organic substances on soils.

ACKNOWLEDGMENTS

This work was partially supported by Bio-Separation Technologies Co., Budapest, Hungary. We are grateful to Miss K. Farkas for her helpful technical work.

REFERENCES

1. Means, J.C., S.G. Wood, J.J. Hassett, and W.L. Banwart. *Environ. Sci. Technol.* 14:1524 (1980).
2. Karickhoff, S.W. *Chemosphere* 10:833 (1981).
3. Karickhoff, S.W., D.S. Brown, and T.A. Scott. *Water Res.* 13:241 (1978).
4. Lambert, S.M. *J. Agric. Food Chem.* 16:340 (1968).
5. Chiou, C.T., L.J. Peters, and V.H. Freed. *Science* 206:831 (1979).
6. Geyer, H., A.G. Kraus, and W. Klein. *Chemosphere* 16:594 (1980).
7. Kenaga, E.E. *Ecotoxicol. Environ. Saf.* 4:26 (1980).
8. Rapaport, R.A. and S. Elsenreich. *J. Environ. Sci. Technol.* 16:163 (1984).
9. Chiou, C.T., P.E. Porter, and D.W. Schmedding. *Environ. Sci. Technol.* 17:227 (1983).
10. Horvath, C. and W. Melander. *J. Chromatogr. Sci.* 15:393 (1977).
11. Vowles, P.D. and R.F.C. Mantoura. *Chemosphere* 16:109 (1987).
12. Hodson, J. and N.A. Williams. *Chemosphere* 17:67 (1988).
13. Szabó, Gy, S.L. Prosser, and R.A. Bulman. *Chemosphere* 21:729 (1990).
14. Hill, J.M. *J. Chromatogr.* 76:455 (1973).
15. Unger, K.K., "Porous Silica" (*J. Chromatogr.* Library Vol. 16) Elsevier, Amsterdam, 1979, 120.
16. Brooke, D.N., A.J. Dobbs, and N.A. Williams. *Ecotoxicol. Environ. Saf.* 11:251 (1986).
17. Miller, M.M., S.P. Wasik, G.-L. Huang, W.-T. Shin, and D. Mackay. *Environ. Sci. Technol.* 19:522 (1985).
18. Braumann, Th., G. Weber, and L.H. Grimme. *J. Chromatogr.* 261:329 (1983).
19. Snyder, L.R., J.W. Dolan, and J.R. Gant. *J. Chromatogr.* 165:3 (1979).

CHAPTER 3

Accelerated Degradation of Soil Insecticides: Comparison of Field Performance and Laboratory Behavior

D. L. Suett and A. A. Jukes

ABSTRACT

Laboratory incubation experiments with aldicarb, carbofuran, and chlorfenvinphos were correlated with field assessments of performance against the cabbage root fly (*Delia radicum*) to show how treatment frequency and site differences influenced the accelerated degradation of their residues and their subsequent biological activity.

I. INTRODUCTION

The behavior of pesticides in soil and their consequent availability to plant and aquatic systems are determined by many factors. Principal among these are physicochemical characteristics of the pesticide, formulation and method of application, and soil and climatic variables. Interactions between these factors lead to the eventual dissipation and/or degradation of all soil-applied organic pesticides. Although the loss of chemical residues may result from physical effects (e.g., volatilization, leaching) and chemical action (e.g., hydrolysis, oxidation), most of the degradation that occurs in soil is caused by microbes.[1] It is therefore essential that effects of microbes are considered when assessing and comparing the observed and potential behavior of pesticides in different soils.

0-87371-616-7/93/$0.00 + $.50

In most instances, microbial activity results in a steady decline of residues so that, after the pesticide has achieved its purpose, excess residues are dissipated. Thus, despite the considerable quantities of pesticides applied each year to agricultural and horticultural soils, there is little evidence of residue accumulation in soil or of carryover into subsequent crops. However, recently there has been increasing evidence that the microbial degradation of pesticide residues may be stimulated to induce unduly rapid degradation of some compounds.[2-4] This phenomenon of accelerated, or enhanced, degradation has been shown to be limiting the stability and hence the biological performance of an increasingly wide range of compounds. Studies in the UK at Horticulture Research International, Wellesbourne (HRI-W) have illustrated some of the potential difficulties that could result from continued development of the phenomenon. However, it also became evident during these studies that the occurrence of accelerated degradation in previously treated field soils was not always associated with inadequate biological performance of the pesticides. Thus an initial laboratory study of soils from brassica-growing farms in different parts of the UK showed greatly increased rates of loss of freshly applied carbofuran from the previously treated field soils compared with similar untreated soils. However, only one of the growers concerned was dissatisfied with the performance of the insecticide against the cabbage root fly (*Delia radicum*).[5] One of the more obvious reasons for such discrepancies is the lack of adequate pest pressure in the field; this can distort the apparent effectiveness of pesticide treatments. Discrepancies may also result from differences in the conditions used for laboratory experiments and those encountered in the field. Often in laboratory experiments, a uniformly incorporated single dose is maintained at a constant temperature and moisture level, whereas in the field nonuniformly dispersed formulations are exposed to fluctuating temperature and moisture. These fluctuations will impose effects on microbial activity which, in the rhizosphere especially, will be influenced further by the distribution of live plant tissues.

Studies of accelerated degradation at HRI-W have therefore attempted, whenever possible, to correlate results of laboratory incubation studies using insecticides with assessments of the activity of insecticides against field populations of insect pests. This chapter describes a study of the influence of the frequency of application of carbofuran to soil on the subsequent laboratory behavior and field performance of aldicarb and carbofuran. Data are also presented from experiments done to compare the behavior and performance of two insecticides, carbofuran and chlorfenvinphos, in soils from eight different fields on the same farm.

II. MATERIALS AND METHODS

The experiments were established at HRI-W on a number of sites, each comprising sandy loam soils of the Wick Series. At each site, initial soil samples were taken to determine the percentage of organic matter, pH, and water content

Table 1. History of Recent Insecticide Treatments and Properties of HRI-W Soils Used in Between-Site Variability Study

Field	Organic matter (%)	pH	H$_2$O at 33 kPa (%)	Treatment history
RP	3.4	6.3	11.2	Nothing since 1983
LMW	5.1	5.3	15.0	Possibly treated in 1982
LMC	5.2	5.7	13.9	Permanent pasture
WG	2.8	6.5	10.3	Nothing since 1980
PG	3.9	6.2	13.9	Nothing since 1981
SW	2.8	5.6	10.7	Treated in 1982
LC	3.3	6.3	11.7	Treated in 1986
HM	3.5	6.6	11.9	Possibly treated in 1982

at an applied pressure of 33 kPa (0.33 bar). Details of the methods used to establish these values have been described.[6] Before sowing, all areas received a base fertilizer application and a seedbed was prepared by harrowing and rolling.

A. Frequency of Carbofuran Application

The experiment comprised 18 beds 15 m long and 1.52 m wide, subdivided into three blocks of 6 beds. All blocks were treated with a 5% active ingredient (a.i.) granular formulation of carbofuran (Yaltox; Bayer) at a dose equivalent to 2.5 kg a.i./ha in May 1986, two blocks received a second application in July, and one block received a third application in September. On each occasion, the granules were mixed with approximately 1 L of dry sand, and this mixture was broadcast evenly by hand and incorporated to a depth of 10 cm with a rotary harrow. A separate six-bed block, which was not treated with insecticide, was established at the same site; and all blocks received similar mechanical treatments during the experiment. Plots were maintained weed-free by hand weeding until late autumn, when paraquat (Gramoxone 100, FBC Ltd.) was applied at a rate equivalent to 600-g a.i. in 500 L water per hectare.

Soil samples were taken in January, April, and July 1987 from all blocks for laboratory incubation studies; and in April 1987 the performance of carbofuran and aldicarb against cabbage root fly was determined as described below.

B. Field-to-Field Variability

A block of six beds, 15 m long and 1.52 m wide, was established in April 1987 in each of the eight fields described in Table 1. Samples were taken for laboratory incubation studies, and the performance of carbofuran and chlorfenvinphos against cabbage root fly was determined as described below.

C. Field Performance Studies

Each of the six-bed areas was treated similarly, with insecticides applied to only the four inner beds and the two outer beds functioning as "guard" beds. Each of the four inner beds comprised three, 15-m single-row plots, 51 cm apart.

Table 2. Granular Insecticide Formulations Used in Field Performance Studies

Insecticide	% a.i.	Supplier	Recommended dose (mg a.i./m) for control of cabbage root fly	Dose-range applied to subplots (mg a.i./m row)
Aldicarb	10	Union Carbide[a]	51	30.2–285
Carbofuran	5	Bayer	62.5	22.4–211
Chlorfenvinphos	10	Shell	70	34.9–329

[a] Now Embetec.

The crop was radish (cv French Breakfast) sown at a mean rate of 64 seeds per meter row. Within each bed, there were one untreated and two treated plots. Continuous exponentially increasing doses of insecticides were applied at sowing by the bow-wave technique,[7] using equipment designed and constructed at HRI-W.[8] The insecticide doses were placed on belt delivery applicators on a Stanhay seed drill[9] using an exponentially grooved trough.[10] The median doses applied to each of the 10 subplots within each plot were calculated from the bulk density of the granules required to fill the trough (Table 2). The direction of the dose gradient was the same for all plots within a bed, but was randomized between blocks.

At harvest, 9–10 weeks after sowing, each 15-m plot was divided into ten 1.4-m subplots after omitting 0.5 m of a row at each end. All radishes in each subplot were lifted, washed, and graded for the presence or absence of damage by cabbage root fly larvae. Dose-response relationships were computed for each insecticide.[11]

D. Laboratory Incubation Studies

Procedures for the preparation and treatment of soils have been described.[6] Samples for incubation studies were taken from the upper 10–12 cm of each area and mixed by sieving. Before incubation, they were air-dried to ca. 50% of their moisture-holding capacities, and their moisture contents were determined accurately. Aliquots of aqueous solutions or suspensions of aldicarb (analytical grade), carbofuran (technical grade), or chlorfenvinphos (24% EC) were applied at a dose equivalent to 25-mg a.i. per kilogram dry soil, followed by further water to adjust the water content of the soil to the appropriate value at 33 kPa. After equilibration and mixing, duplicate portions were transferred to wide-necked bottles, covered loosely with aluminium foil and incubated in the dark at 15°C. Soil moisture contents were maintained by twice weekly additions of water; and at intervals during the subsequent eight weeks, 30-g samples were removed and stored at − 15°C until analysis.

E. Analytical Methods

Residues of aldicarb and carbofuran were analyzed by high-performance liquid chromatography (HPLC)[6] and residues of chlorfenvinphos by gas-liquid chro-

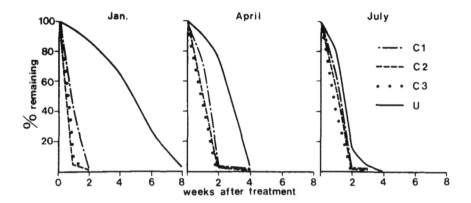

Figure 1. The decline in 1987, under controlled laboratory conditions, of freshly applied carbofuran in soils treated with the insecticide in 1986. C1, C2, C3 — treated 1, 2, or 3 times; U — no previous treatment.

matography (GLC).[12] Analytical efficiencies exceeded 95%, and results were not corrected for analytical losses.

III. RESULTS

A. Frequency of Carbofuran Application

The behavior of freshly applied carbofuran in laboratory incubation studies with soils from the four areas sampled in January, April, and June 1987 is shown in Figure 1. Results are expressed for each sampling occasion as the percentage of amounts of carbofuran present in the samples taken immediately prior to incubation.

These studies showed that rates of carbofuran degradation were similarly rapid in all the previously treated soils and changed little from January to July. However, the stability of carbofuran in the previously untreated soil changed markedly over this period. Thus the time for initial 50% loss from the previously treated soils was always 6–10 days whereas in the previously untreated soil it declined from 5 weeks in the sample taken in January to 3 and <2 weeks in the April and July samples, respectively. With the July samples, this resulted in nearly identical rates of degradation in the untreated soil and in all three treated soils.

In the field performance studies established on these soils in April 1987, infestations of cabbage root fly were moderately severe, with 71 (70–73) and 57% of the untreated radish damaged in the treated and untreated soils, respectively. The values of the intercepts (α) and slopes (β) of the dose-response models for the log-dose applications of both insecticides to the different soils are given in Table 3.

The extent to which the 1986 carbofuran treatments affected the performances of aldicarb and carbofuran is shown in Figure 2. Although the performance of

Table 3. Intercepts (α) and Slopes (β) of the Linear Dose-Response Models for Insecticide Treatments Applied to Soils C1–C3, CU in 1987

Insecticide	Parameters	Parameters ($\pm$ SE) for models			
		Soil C1	Soil C2	Soil C3	Soil CU
Aldicarb	α	1.51 $\pm$ 0.33	1.96 $\pm$ 0.36	1.82 $\pm$ 0.43	1.89 $\pm$ 0.39
	β	-1.31 ± 0.18	-1.55 ± 0.20	-1.52 ± 0.23	-1.47 ± 0.22
Carbofuran	α	[a]	0.60 $\pm$ 0.25	[a]	1.19 $\pm$ 0.75
	β	[a]	-0.35 ± 0.14	[a]	-1.70 ± 0.46

[a] Slope of regression line not significantly different from zero ($p = 0.05$).

aldicarb was similar in all soils, the recommended dose (51-mg a.i. per meter) reduced the numbers of larvae by only 50%. In contrast, the performance of carbofuran was substantially lower in all the pretreated soils than in the untreated soil, where the recommended dose (62.5-mg a.i. per meter) reduced the numbers of larvae by 84%. In two of the pretreated soils, the slopes of the carbofuran dose-response regressions were not significantly different ($p = 0.05$) from zero; and in the other, treated with carbofuran on two occasions, only the three highest-dose subplots showed a limited dose response. Thus the performance of the largest dose of 211-mg a.i. per meter in all the pretreated soils was poorer than that of the smallest dose, 22-mg a.i. per meter, in previously untreated soil.

B. Field-to-Field Variability

Incubation studies with the eight HRI-W soils showed marked differences in the behavior of carbofuran and chlorfenvinphos (Figure 3). Chlorfenvinphos was consistently more persistent, with 50–80% of the applied dose remaining after 8 weeks. In contrast, six of the soils contained <5% of the applied dose of carbofuran after this time. Both insecticides persistend longest in LMW, which lost only 20% during the incubation; and carbofuran was only slightly less persistent than chlorfenvinphos in LMC, where 45 and 60%, respectively, remained by the end of the study. The major difference in behavior between insecticides and soils was the duration of the initial "lag" phase. In all the soils, chlorfenvinphos declined steadily and relatively slowly throughout the 8 weeks, a pattern of loss evident with carbofuran only in LMW. In all the other soils, the duration of this carbofuran "lag" phase ranged from 6 weeks (LMC, SW) to 1 week (LC) or less (RP), after which there was a rapid and similar acceleration of loss.

Values of the intercepts (α) and slopes (β) of the dose-response models from the field performance experiment, undertaken at the same time as the incubation studies, are given in Table 4 together with computed estimates of the larval mortalities at doses of 60- and 90-mg a.i. per meter row. The mean level of cabbage root fly damage to the untreated crop was 66% and ranged from 41 (WG) to 84% (RP). Chlorfenvinphos performed consistently well at all sites, a dose of 60-mg a.i. per meter reducing larval numbers by 89–95%. Nevertheless, at six of the sites carbofuran was always more effective, the recommended dose of 62.5-mg a.i. per meter row reducing the numbers of larvae by 96.2–99.3%

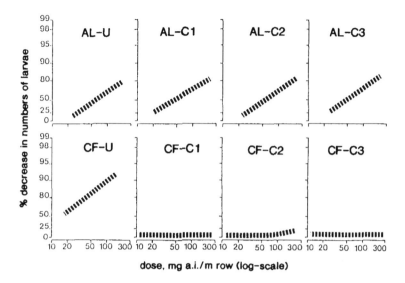

Figure 2. The relationship between the doses of aldicarb (AL) and carbofuran (CF) applied to radish in April 1987 and the estimated percentage decrease in the numbers of cabbage root fly larvae in soil treated with carbofuran in 1986. C1, C2, C3 — treated 1, 2, or 3 times; U — no previous treatment.

compared with 89.6–94.9% reduction with 90-mg chlorfenvinphos per meter. In contrast, the performance of carbofuran in the RP and LC soils was substantially worse than that of chlorfenvinphos, the recommended doses achieving only 72 and 54% reductions, respectively, in the numbers of larvae.

IV. DISCUSSION

A. Frequency of Treatment

In a previous study, Suett[13] showed that multiple applications of carbofuran reduced its stability and biological performance considerably in the following year. The present study of treatment frequency confirmed this observation and also showed that a single application of the recommended dose of carbofuran is sufficient to accelerate the degradation of a subsequent field application more than 12 months later. Furthermore, this accelerated degradation reduced the biological activity of carbofuran over a wide dose range and led to inadequate control of cabbage root fly, a major soil pest in the UK. For the practical implications of this observation to be fully understood, it is essential that the extent to which the phenomenon of accelerated degradation exists and persists is established as quickly as possible.

It was significant also that the stability and performance of aldicarb were similar in all these soils. A similar observation was reported following five broadcast applications of carbofuran[13] although a subsequent study showed that

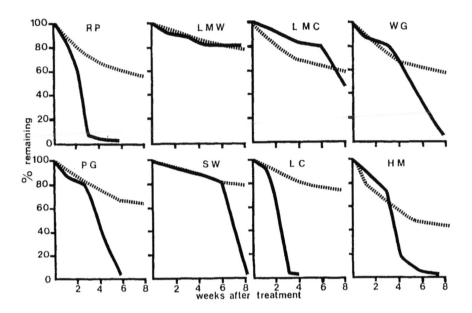

Figure 3. The decline under controlled laboratory conditions of ——— carbofuran and – – – chlorfenvinphos in soils from eight sites at HRI-W.

a single band application of 171-mg carbofuran per meter stimulated the accelerated degradation of aldicarb,[6] suggesting that this particular cross-enhancement may be influenced more by the dose than by the frequency of treatment.

The decline in stability of carbofuran in the previously untreated soil from January to July seems most likely to have resulted from a progressive increase in the degradative capacity of the untreated soil rather than from any seasonal flux in general soil microbial activity, as rates of loss from the three previously treated soils remained unchanged during this period. A similar modification of previously untreated soils was reported by Suett and Jukes,[14] who found that soils from hedgerows and ditches surrounding fields used previously to grow carrots degraded phorate at similar accelerated rates to those observed in the cultivated soils. In the present experiment, the untreated area was approximately 50 m from the treated plots; and with evidence that the ability to degrade carbofuran rapidly can be induced by transferring as little as 0.25% of an "active" soil,[15] the possibility of cross-contamination cannot be discounted. The observation reinforces the advice that untreated sites should be located, defined, and maintained with diligence.[6]

Nevertheless, despite its increasingly rapid degradation in laboratory studies with the previously untreated soil, carbofuran remained highly effective against cabbage root fly in this soil following application in late April. The results illustrate the importance of supporting laboratory incubation data with field evaluations of biological efficacy.

Table 4. Intercepts (α) and Slopes (β) of Linear Dose-Response Models, Together with Estimates of Larval Mortalities at Two-Dose Levels, Following Application of Carbofuran and Chlorfenvinphos at Eight Sites

	Chlorfenvinphos				Carbofuran			
	Parameters (±SE)		Estimated mortality at (mg a.i./m)		Parameters (±SE)		Estimated % mortality at (mg a.i./m)	
Site	α	β	60	90	α	β	60	90
RP	2.24 ± 0.59	−2.49 ± 0.36	88.7	92.7	−0.13 ± 0.32	−0.65 ± 0.19	72.2	75.2
LMW	3.95 ± 0.56	−3.55 ± 0.36	90.8	95.1	1.22 ± 0.87	−2.52 ± 0.60	96.2	97.5
LMC	2.72 ± 0.70	−3.01 ± 0.45	92.9	95.8	1.91 ± 0.86	−3.25 ± 0.61	97.9	98.8
WG	3.28 ± 0.63	−3.11 ± 0.40	89.6	94.0	2.33 ± 0.87	−3.51 ± 0.63	98.0	98.9
PG	4.18 ± 0.62	−3.63 ± 0.41	90.4	94.9	3.96 ± 0.76	−4.61 ± 0.56	98.6	99.4
SW	3.74 ± 0.62	−3.77 ± 0.41	94.9	97.4	4.17 ± 0.94	−5.11 ± 0.71	99.3	99.7
LC	2.37 ± 0.71	−2.76 ± 0.45	92.0	95.1	0.48 ± 0.37	−0.71 ± 0.22	53.8	59.2
HM	3.15 ± 0.70	−3.13 ± 0.45	91.1	94.9	2.21 ± 0.66	−3.14 ± 0.46	96.6	98.0

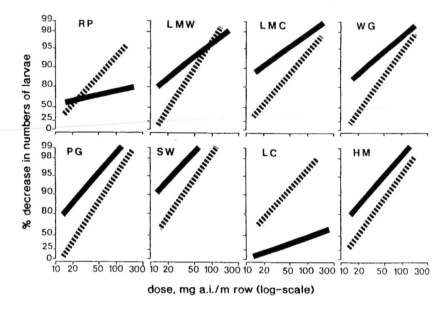

Figure 4. The relationship between the doses of ——— carbofuran and – – – chlorfenvinphos
applied to radish and the estimated percentage decrease in the numbers of cabbage root fly larvae in soils from eight sites at HRI-W.

B. Field-to-Field Variability

The reasons for marked differences in the behavior and performance of carbofuran at the eight sites are not known. Although the accelerated loss in soil
LC seems likely to have been induced by the single application of the recommended dose of carbofuran 12 months earlier, similarly rapid loss occurred in
soil RP which had not received any insecticide treatment for at least 5 years.
Furthermore, although all the other soils had remained untreated with insecticides
for at least several years, there were significant differences in the duration of
the initial lag phases. With such large differences occurring between different
fields on the same farm, it is evident that much is yet to be learned about the
interactions of soil properties with cultural practices and their impact on the
behavior of pesticides in soils.

The results of laboratory and field experiments associated with this study
emphasize further the merits of an integrated approach. There seems little doubt
that extrapolation from the incubation studies (Figure 3) would have enabled the
satisfactory performance of chlorfenvinphos at all sites, and of carbofuran in
soils LMW, LMC, and SW to be predicted with some confidence. It would also
have indicated correctly that crops in soils RP and LC were likely to be protected
least. However, it would have been difficult to predict from those studies alone
the extent to which carbofuran would perform in soils PG, HM, and probably
WG. It should be recognized also that radish is a relatively short-term crop and

that control of cabbage root fly on this crop will not necessarily guarantee adequate protection of brassica crops which remain susceptible for longer periods (Figure 4).

REFERENCES

1. Graham-Bryce, I.J. in *The Chemistry of Soil Processes,* D.J. Greenland and M.H.B. Hayes, Eds. (Chichester: Wiley-Interscience, 1981), pp. 621–670.
2. Kaufman, D.D., J. Katan, D.F. Edwards, and E.G. Jordan. in *Agricultural Chemicals of the Future,* J.L. Hilton, Ed. (Totowa, NJ: Rowman & Allanheld, 1984), pp. 437–451.
3. Roeth, F.W. *Rev. Weed Sci.* 2:45–65 (1986).
4. Suett, D.L. and A. Walker. *Aspects Appl. Biol.* 17:213–222 (1988).
5. Suett, D.L. *Crop Prot.* 5:165–169 (1986).
6. Suett, D.L. and A.A. Jukes. *Crop Prot.* 7:145–152 (1988).
7. Makepeace, R.J. *Proc. 3rd Br. Crop Prot. Conf.* 1:389–395 (1965).
8. Thompson, A.R. *Meded. Fac. Landbouwwet. Rijksuniv. Gent* 4:909–918 (1984).
9. Thompson, A.R., A.L. Percivall, G.H. Edmonds, and G.R. Lickorish. *Ann. Appl. Biol.* 102:511–521 (1983).
10. Wheatley, G.A. *Proc. 6th Br. Insecticide Fungicide Conf.* 2:386–395 (1971).
11. Phelps, K. and A.R. Thompson. *Ann. Appl. Biol.* 103:191–200 (1983).
12. Suett, D.L. *Pestic. Sci.* 2:105–112 (1971).
13. Suett, D.L. *Crop Prot.* 6:371–378 (1987).
14. Suett, D.L. and A.A. Jukes. *Toxicol. Environ. Chem.* 18:37–49 (1988).
15. Harris, C.R., R.A. Chapman, C. Harris, and C.M. Tu. *J. Environ. Sci. Health,* B19:1–11 (1984).

CHAPTER 4

Adsorption of 2,4-D on Organoclays

María C. Hermosín and Juan Cornejo

ABSTRACT

The adsorption capacity of decylammonium-montmorillonite (C10-M) and decylammonium-vermiculite (C10-V) for the anionic pesticide 2,4-dichloro-phenoxyacetic acid (2,4-D) has been measured from the adsorption isotherms and compared with those of the inorganic clays (Na$^+$-M and Na$^+$-V). 2,4-D adsorption capacity was larger for C10-V (high layer charge mineral) than for C10-M (low layer charge mineral). The uptake of 2,4-D by organoclays was probably due to lyophilic interactions between the alkyl chains of the adsorbents and the aromatic ring of the herbicide, this adsorption being similar to a partitioning process between water and the organic phase of organoclays. The organic phase of C10-M behaved similarly to the soil organic matter, but the organic phase of C10-V (even being the same) adsorbed five times more 2,4-D than soil organic matter did. The higher adsorption capacity of C10-V, as compared with C10-M, was related to the different arrangement of organic cations in the interlayer spaces of these clays, which was determined by their layer charge. Successive desorption results showed also that 2,4-D binding to C10-V was stronger than to C10-M. The results of this study suggest the possibility of using these organoclays for removing 2,4-D from contaminated waters (C10-V) and for a possible slow-release formulation (C10-M).

0-87371-616-7/93/$0.00 + $.50

I. INTRODUCTION

The clay-organic interactions have been extensively studied; and the organoclay complexes have been recently shown to be good sorbents for nonpolar organic pollutants such as benzene, toluene, and xylene[1,2] or for polar organic pollutants such as phenols[3-5a] or organophosphorous pesticides.[6] The organoclay complexes have been shown also to promote the degradation of a cationic pesticide such as chlordimeform.[7] Even the formation of such synthetic organoclay complexes has been proposed as a medium for decreasing the mobility of organic pollutants in soils.[8]

The replacement of the initial inorganic exchangeable cations of clays by organic cations, such as alkylammonium, has been shown to produce a great improvement in the sorption capacity of clays for organic pollutants.[2-4] This sorption capacity was directly related to the size of the organic cation present in the clay complex[4] and inversely related to the layer charge of clay.[2]

Thus the organoclays have been developed as a tool to decrease the contamination of soils and waters. The present investigation was conducted to check the adsorption capacity of such organoclays for the acid herbicide 2,4-D which due to its anionic character has high mobility in soils; and thus it is a potential pollutant for groundwaters because it has occurred with nitrates. In the present work two phyllosilicates with different layer charges were used to prepare organoclay complexes by saturation with decylammonium ions. The adsorption capacity was studied by 2,4-D adsorption isotherms, and desorption by successive dilution was used to check the reversibility of the adsorption process.

II. MATERIALS AND METHODS

The clays used in this study were the SWy-1 reference sample of the Clay Minerals Repository (M) and the Santa Olalla-Spain vermiculite (V). The organoclay complexes were prepared by treating the clay with 0.5 N decylammonium chloride solution several times and shaking for 24 hr. The vermiculite sample was heated for 8 hr at 60°C to promote the exchange reaction. After that the clays were washed with distilled water until chloride free, dried at room temperature, and gently hand ground. The properties of these samples are summarized on Table 1.

Table 1. Properties of Untreated and Decylammonium-Treated Clays

Sample	Cation exchange capacity (meq/100 g)	Organic carbon (g/100 g)	Organocation (meq/100 g)	d_{001} (Å)	N$_2$ Surface area (m²/g)
Na+-M	76.4	—	—	15.4	60.0
Na+-V	130.0	—	—	14.2	3.0
C10-M	—	10.3	78.9	13.6	10.0
C10-V	—	17.5	142.0	21.6	3.0

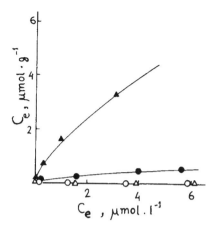

Figure 1. 2,4-D adsorption isotherms on: (○)Na$^+$-M, (△)Na$^+$-V, (●)C$_{10}$-M, and (▲)C$_{10}$-V.

2,4-D adsorption isotherms were carried out by the batch equilibration technique using five initial concentrations between 1 and 50 μM in CaCl$_2$ 0.01 M. The initial solutions were labeled with [^{14}C]-2,4-D. Duplicates of 0.1 g of solid were added to 5 mL of 2,4-D-labeled solutions, shaking 24 hr (previously it was thought that 15 hr was enough time to reach the equilibrium) and centrifuged at 20,000 rpm. The supernatants were decanted and analyzed by liquid scintillation counting. The amount of 2,4-D adsorbed was assumed to be the difference between initial and final concentration of the herbicide in solution.

The desorption by successive dilution was carried out after adsorption: 3 mL of the 2,4-D equilibrium solution was removed; 3 mL of the solvent (CaCl$_2$ 0.01 M) was added, shaken for 24 hr, and centrifuged; then the supernatant was analyzed. This was repeated three times for duplicate samples.

III. RESULTS AND DISCUSSION

The adsorption isotherms of 2,4-D on the untreated and decylammonium-treated clays, used as adsorbents, are shown in Figure 1. 2,4-D adsorption on untreated clays were almost negligible (as had been reported earlier[9,10]) because the negative charge of these clays repelled the acidic or anionic form of these herbicides, but preferentially adsorbed water in such a way that for some curve points negative adsorption was found. 2,4-D adsorption on C10-M and C10-V was higher than that corresponding to the untreated samples. The shapes of these isotherms showed L and C characteristics,[11] respectively, indicating high affinity of the solute for these adsorbents and even possible penetration of the solute in intracrystalline spaces of these sorbents.

2,4-D has been shown to be highly adsorbed by charcoal,[10] and its adsorption by soils and sediments was closely related to the organic matter content.[12-14] The

Table 2. Freundlich 2,4-D Adsorption Parameters for
the Adsorbents Studied

Adsorbent	K_f (μmol/g)	nf	K_{oc} (μmol/g)	Soil K_{oc}
Na$^+$-M	0.004	0.61	—	—
Na$^+$-V	0.004	0.73	—	—
C10-M	0.067	1.10	0.68	0.01–0.65[a]
C10-V	1.668	0.99	8.60	0.14–2.21[b]

[a] Hermosín and Cornejo.[12]
[b] Moreale and Van Bladel.[14]

higher adsorption capacity for 2,4-D showed by organoclays was related to their organic carbon content; thus the organic cation in the interlamellar spacings acts as the organic matter did in soils. According to Boyd et al.[3,4] the organic cation in the interlayer of clays acted as a partition medium for the solute, because of the lyophilic nature of 2,4-D.

The adsorption isotherm data were assayed to fit the Freundlich equation[15] for which the parameters are shown on Table 2. Effectively the substitution of inorganic exchangeable cations by organic decylammonium ions highly increased the adsorption capacity (K_f) and intensity (nf) for 2,4-D. The nf parameter for 2,4-D adsorption on organoclays was very close to the unit indicating an almost constant partition of the solute between the sorbent and the solvent. The adsorption capacity of C10-V was found to be much higher than that of C10-M, which can be related to the organic carbon content of these clays (Table 1); the organic phase of these adsorbents acted as organic matter did in soil, as has been suggested above. However, the difference in 2,4-D adsorption capacity was much greater than the difference for organic carbon content between C10-M and C10-V.

We can calculate the adsorption capacity of these organoclays on the basis of their organic carbon content (K_{oc}) as done for soils or sediments. K_{oc} values for 2,4-D adsorption on organoclays studied are reported in Table 2 besides some values found in the literature. The K_{oc} value for C10-M was similar to those reported for soils; however, the K_{oc} value for C10-V was higher. This implies that the same organic unit (decylammonium ions) in vermiculite had higher affinity for 2,4-D than that in montmorillonite. This could be related to the arrangement of organic cations in the interlayer space of these clays which determines the size of the organic layer adsorbing pesticide by lyophilic bonds.

For C10-M the organic cations lie with the alkyl chain parallel to the silicate layer, because of the equivalent area (specific area per charge unit) was larger than that occupied by the flat organic cation. This disposition, as shown in Figure 2a, displayed a basal spacing of 13.6 Å (Table 1) which implies an external organic layer thickness of 4.0 Å besides an inorganic one of 9.6 Å. This organic layer thickness was almost equal to that of the 2,4-D molecule, with the ring parallel to the silicate layer which probably should be adsorbed this way on the external organic surface, as shown in Figure 2a.

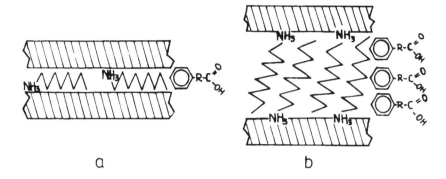

Figure 2. Interlayer organic cations and 2,4-D adsorbed molecule arrangement on (a) C_{10}-M and (b) C_{10}-V.

Table 3. Percentage of Desorption by Successive Dilution

Sample	Number of dilutions			Total desorbed
	1	2	3	
C10-M				
Cs = 2.5 μmol/g	30.0	30.0	27.0	67.0
C10-M				
Cs = 4.5 μmol/g	4.0	5.0	4.0	12.0

For the C10-V sample due to the high layer charge of this mineral, the organic cations have to arrange with their alkyl chains tilting 55° to the silicate layer because in such disposition the area occupied by the organic cation was similar to the equivalent area of vermiculite mineral. This organic cation arrangement (Figure 2b) produced a basal spacing of 21.6 Å for C10-V composed by an external organic layer thickness of 12.0 Å and an inorganic one of 9.6 Å. Thus the external organic surface for establishing lyophilic bonds between the organic cation and the organic herbicide was larger than that for C10-M, as shown in Figure 2b; hence the adsorption capacity of 2,4-D was found to be higher for C10-V than for C10-M.

The contradiction found between the present results showing higher organic adsorption capacity for high charge vermiculite and those reported by Lee et al.[2] showing lower organic adsorption capacity for high charge smectite, was due to the Lee work involving the disposition of interlayer organic cations which was the same for high and for low charge montmorillonites. For this reason both types of samples displayed the same basal spacing; and therefore the possible interlamellar adsorption of the organic pollutant was more difficult for the high charge organic montmorillonite.

The results of desorption by successive dilution are shown in Table 3 as percentage of desorption of the initial amount of 2,4-D adsorbed. These results showed that 2,4-D desorption from C10-M was much easier than from C10-V;

this was because after three successive desorptions, 67% of the 2,4-D initially adsorbed on C10-M whereas only 1% had been desorbed from C10-V. These data suggested that the binding of 2,4-D molecules was stronger in C10-V than in C10-M.

The results found in this study show that decylammonium clays, especially the high charge vermiculite, were very effective for removing 2,4-D from water and thus may be used as purifying agents for 2,4-D-contaminated waters, as well as in places where the mobility of 2,4-D needs to be decreased. Also C10-M, due to its facility for desorbing 2,4-D, could be used as support for the slow release formulation of 2,4-D or other anionic or acidic herbicides.

ACKNOWLEDGMENT

This work has been supported by the Junta de Andalucía and the DGICYT (Project No. NAT91-1336).

REFERENCES

1. Lee, J.F., M.M. Mortland, S.A. Boyd, and C.T. Chiou. "Shape-Selective Adsorption of Aromatic Molecules from Water by Tetramethyl Ammonium Smectite." *J. Chem. Soc. Faraday Trans. 1* 85:2953–2962 (1989).
2. Lee, J.F., M.M. Mortland, C.T. Chiou, D.E. Kile, and S.A. Boyd. "Adsorption of Benzene, Toluene and Xylene by Two Tetramethylammonium-smectites Having Different Layer Charge Densities," *Clays Clay Miner.* 38:113–120 (1990).
3. Boyd, S.A., M.M. Mortland, and C.T. Chiou. "Sorption Characteristics of Organic Compounds on Hexadecyltrimethylammonium-Smectite," *Soil Sci. Soc. Am. J.* 52:652–657 (1988).
4. Boyd, S.A., S. Shaobai, J.F. Lee, and M.M. Mortland. "Pentachlorophenol Sorption by Organo-Clays," *Clays Clay Miner.* 36:120–130 (1988).
5. Srinavasan, K.R. and H.S. Fogler. "Use of Inorgano-Organo Clays in the Removal of Priority Pollutants from Industrial Wastewaters: Structural Aspects," *Clays Clay Miner.* 38:277–286 (1990).
5a. Srinavasan, K.R. and H.S. Fogler. "Use of Inorgano-Organo Clays in the Removal of Priority Pollutants from Industrial Wastewaters: Adsorption of Benzo(a)-Pyrene and Chlorophenols from Aqueous Solutions," *Clays Clay Miner.* 38:287–293 (1990).
6. Sanchez-Camazano, M. and M.J. Sanchez-Martin. "Sorption Capacity Modification of Smectites for Organophosphorous Pesticides," *5th Int. Symp. Environmental Pollution and Its Impact on Life in the Mediterranean Region*, Blanes, Spain, 1989, Abstract Book, 19.
7. Morillo, E., J.L. Perez Rodriguez, and M.C. Hermosin. "Interaction of Chlordimeform with a Vermiculite-Decylammonium Complex in Aqueous and Butanol Solutions," *Miner. Petrogr. Acta* 29-A:155–162 (1985).
8. Boyd, S.A., J.F. Lee, and M.M. Mortland. "Attenuating Organic Contaminant Mobility by Soil Modification," *Nature* 333:345–347 (1988).

9. Ogram, A.V., R.E. Jessup, L.T. Ou, and P.S.C. Rao. "Effects of Sorption on Biological Degradation Rates of 2,4-Dichlorophenoxy Acetic Acid in Soils," *Appl. Environ. Microbiol.* 1985:582–587 (1985).

10. Weber, J.B., P.W. Perry, and R.P. Upchurch. "The Influence of Temperature on the Adsorption of Paraquat, Diquat, 2,4-D and Prometrone by Clays, Charcoal and Anion-Exchange Resin," *Soil Sci. Soc. Am. Proc.* 29:678–688 (1965).

11. Giles, C.H., T.H. MacEwan, S.N. Nakhwa, and D. Smith. "Studies in Adsorption. Part XI. A System of Classification of Solution Adsorption Isotherms and Its Use in Diagnosis of Adsorption Mechanisms and in Measurements of Specific Surface Area," *J. Chem. Soc.* 3973–3993 (1960).

12. Hermosin, M.C. and J. Cornejo. "Soil Adsorption of 2,4-D as Affected by the Clay Mineralogy," *Toxicol. Environ. Chem.* 31–32:69–77 (1991).

13. Lokke, H. "Sorption of Selected Organic Pollutants in Danish Soils," *Ecotoxicol. Environ. Saf.* 8:395–409 (1984).

14. Moreale, A. and R. Van Bladel. "Fate of 2,4-D in Belgium Soils," *J. Environ. Qual.* 9:627–630 (1980).

15. Hermosin, M.C. and J. Cornejo. "Maleic Hydrazide Adsorption by Soils and the Use of the Mole Fraction form of the Freundlich Equation," *Soil Sci.* 14:453–456 (1987).

CHAPTER 5

Competitive Adsorption of 2,4-D and Phosphate in Soils

L. Madrid, E. Morillo, and E. Diaz-Barrientos

I. INTRODUCTION

Addition of phosphate to soils is a common agricultural practice. As phosphate anions are known to adsorb strongly on many soil minerals, the presence of important amounts of phosphorus fertilizers is likely to cause a noticeable decrease in the adsorbing properties of soils with respect to other anions, e.g., anionic pesticides. This chapter studies the competitive adsorption of the anionic herbicide 2,4-dichlorophenoxyacetic acid (2,4-D) and phosphate on various soils in southwestern Spain. This herbicide was chosen because it was one of those recommended in EEC Directive 79/831 for the ring test on adsorption-desorption of chemicals in soils, conducted in 1989 with the participation of the authors of this chapter together with other European laboratories.[1]

Because many adsorbing properties of soils can be related with variable charge surfaces present in the clay fraction, in a previous study the competition was studied on an iron oxyhydroxide of a large specific surface area, lepidocrocite, and it was concluded that the decreasing effect of phosphate on 2,4-D adsorption could be explained in terms of changes in the surface electrical potential, suggesting that 2,4-D retention is mainly due to electrostatic forces. The order of addition of both adsorbates did not affect final distribution of the herbicide, but P adsorption was retarded by the presence of 2,4-D previously adsorbed.[2]

0-87371-616-7/93/$0.00 + $.50

Table 1. Some Properties of the Soils Used

Soil	pH	Clay (%)	CO_3 (%)	o.m.[a] (%)	Clay minerals and % clay fraction
I	7.5	68	16	2.2	Illite (75%)
II	5.3	27	0	0.3	Kaolinite (85%)
III	7.6	51	16	3.3	Illite (46%)
					Smectite (28%)
					Kaolinite (26%)

[a] Other material.

II. MATERIALS AND METHODS

The iron oxide used has been extensively used as an adsorbent, and its properties have been described previously.[3,4] Two clay minerals, montmorillonite (source — Clay Minerals Repository SAZ-1 from Arizona) and kaolinite (KGa-1 from Georgia), were also used in some experiments.

Three soils from southwestern Spain were chosen, and the main characteristics are given in Table 1.

The experiments were done in 50 mL polypropylene centrifuge tubes, by mixing a given amount of solid with 0.01 M NaCl (in some cases KCl) solutions containing various concentrations of P and/or 2,4-D. The P compound was KH_2PO_4 or NaH_2PO_4, and the herbicide was in acid form and contained 2 × 10^{-4} mCi of ^{14}C-labeled 2,4-D (Sigma Chem. Co.) per milligram of herbicide. In some of the experiments with pure minerals different pH values were studied; and pH adjustments were carried out by adding small, exactly measured volumes of 0.1 M HCl or NaOH with the final pH taken as the pH of the experiments.

Final P concentrations were measured spectrophotometrically by the method of Murphy and Riley,[5] and the initial and final radioactivity of the solutions were measured by liquid scintillation counting. The time of counting was chosen to obtain a confidence level of better than ±1%.

III. RESULTS AND DISCUSSION

Figure 1 shows some examples of isotherms of P-free 2,4-D at various pH values, taken from Madrid and Diaz-Barrientos.[2] A strong pH dependence is observed, as expected in the case of variable charge surfaces; and a positive concavity suggests favorable Van der Waals interactions between adsorbed ions, as proposed by Kavanagh et al.[6] for other adsorbates related with the 2,4-D. For a given pH value (pH 3, near the value for maximum adsorption, pH = pK = 2.73), a drastic reduction in the adsorption levels is found when P is present (Figure 2); as the amount of P increases the concavity progressively disappears or even seems to change to negative. Madrid and Diaz-Barrientos[2] showed that this behavior could be explained by means of the Stern model.

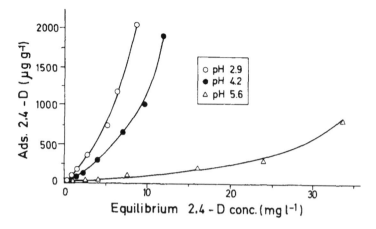

Figure 1. Adsorption isotherms of 2,4-D by iron oxide at various pH values. [From L. Madrid and E. Diaz-Barrientos, *Aust. J. Soil Res.* 29:15–23 (1991).]

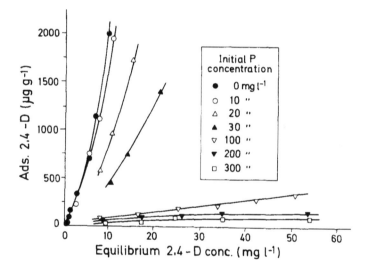

Figure 2. Adsorption isotherms of 2,4-D by iron oxide at pH 3 in the presence of various P concentrations. [From L. Madrid and E. Diaz-Barrientos, *Aust. J. Soil Res.* 29:15–23 (1991).]

The behavior of a mainly constant charge mineral seems to be different from that described for the iron oxide. Table 2 shows two examples of adsorption of 2,4-D by montmorillonite at different pH values. The pH dependence is much less evident and the amounts adsorbed are quite small, so that the influence of the presence of P was difficult to notice; and therefore the results are not shown here. An experiment with larger solid/solution ratios designed to obtain larger

Table 2. pH-Dependence of 2,4-D Adsorption by Montmorillonite

Initial conc.	Final conc. pH 7.6 (mg/L)	Final conc. pH 4.6
5	4.9	4.8
10	9.8	9.5
20	19.9	19.3
30	29.8	29.3
60	59.4	58.4

Note: Experiments with 0.2-g montmorillonite in 10 mL 0.01 *M* NaCl solution.

Table 3. Negative Adsorption of 2,4-D by Montmorillonite in Two Background Electrolytes

Initial 2,4-D conc. (mg/L)	0.01 *M* NaCl		0.01 *M* KCl	
	Final conc. (mg/L)	d_{excl} (A)	Final conc. (mg/L)	d_{excl} (A)
10	10.4	41	10.1	10
20	20.3	17	20.2	11
30	30.4	15	30.3	10
50	50.6	12	50.4	8
80	81.8	22	81.8	22

Note: Experiments with 1-g montmorillonite in 10 mL solution, pH 7.6.

adsorption was unsuccessful because the final solution concentrations of 2,4-D were higher than the initial ones, probably due to anionic exclusion as a consequence of the larger concentration of negatively charged surfaces (Table 3). It can be easily calculated that this behavior corresponds to an anion exclusion region of about 15 Å on the clay surface, which is consistent with data of anion exclusion in the literature.

The possibility of a decrease in effective negative charge and consequently in anionic exclusion due to specific adsorption of potassium, which has been suggested by some authors, was tested by repeating the experiment just described using 0.01 *M* KCl as the background electrolyte. The data suggest a slight decrease in exclusion thickness, but there is not enough evidence; and, in any case, the effect is not important for this clay mineral.

Figure 3 shows the adsorption isotherms of 2,4-D by soils II and III. Soil I, a saline soil with illite as the main clay component, gave final concentrations slightly larger than the initial values. This result is consistent with a large density of negative charge because illite in a sodium-rich (saline) environment is likely to have released part of its interlayer potassium, exposing part of its permanent charge to the adsorbing 2,4-D anions. The isotherms, obtained at the original pH of the soil, showed a positive concavity as those for the iron oxide and in contrast with the isotherms of P adsorption by the same soils (Figure 4). Presence of various P concentrations in soil III gave a detectable increase in 2,4-D adsorption with little sensitivity to changes in the amount of P (Figure 5). On the

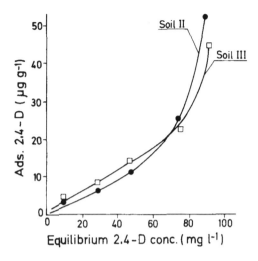

Figure 3. Adsorption isotherms of 2,4-D by soils II and III.

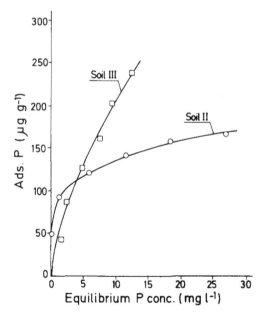

Figure 4. Adsorption isotherms of phosphate by soils II and III.

other hand, soil II showed little changes when P was present (Figure 6). When the experiments are repeated with Na phosphate, soil II showed a decrease in 2,4-D adsorption as expected from the behavior of the iron oxide (Figure 7). Soil III showed a similar, less marked behavior (Figure 8), but in both cases the effect was small.

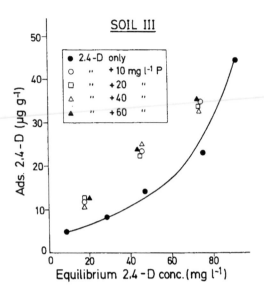

Figure 5. Effect of various concentrations of KH$_2$PO$_4$ on the adsorption isotherm of 2,4-D by soil III.

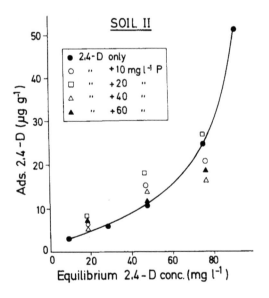

Figure 6. Effect of various concentrations of KH$_2$PO$_4$ on the adsorption isotherm of 2,4-D by soil II.

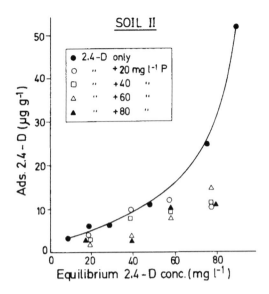

Figure 7. Effect of various concentrations of NaH$_2$PO$_4$ on the adsorption isotherm of 2,4-D by soil II.

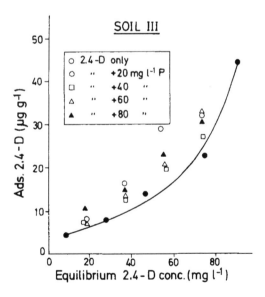

Figure 8. Effect of various concentrations of NaH$_2$PO$_4$ on the adsorption isotherm of 2,4-D by soil III.

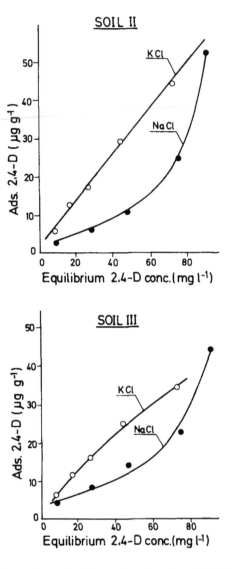

Figure 9. Effect of the background cation on the adsorption of 2,4-D by soils II and III.

The behavior observed suggested the obtaining of isotherms in KCl as the background electrolyte. A spectacular increase in 2,4-D adsorption was observed in both soils (Figure 9). The reason for this effect of the presence of K in solution on the adsorption of the herbicide is not clear, and some additional experiments currently are being conducted to clarify this point. Data obtained with kaolinite (not shown) suggest that no differences exist for this mineral when different alkaline cations are used in the electrolyte.

REFERENCES

1. Herrmann, M. Evaluation of the EEC Laboratory Ring-test, "Absorption/Desorption of Chemicals in Soil." In: Euro-Soils, Identification, Collection, Treatment, Characterization. JRC, Commission of the European Communities, Ispra, 1992.
2. Madrid, L. and E. Diaz-Barrientos. *Aust. J. Soil Res.* 29:15–23 (1991).
3. Cabrera, F., Arambarri, P., Madrid, L. and Toca, and C.G. *Geoderma* 26:203–216 (1981).
4. Madrid, L. and Arambarri, P., *J. Soil Sci.* 36:523–530 (1985).
5. Murphy, J. and J.P. Riley. *Anal. Chim. Acta* 27:31–36 (1962).
6. Kavanagh, B.V., A.M. Posner, and J.P. Quirk. *J. Soil Sci.* 31:33–39 (1980).

CHAPTER 6

Adsorption of Two Weak Acids on Goethite

J. C. Dur, R. K. Khandal, and M. Terce

ABSTRACT

Adsorption characteristics of 2,4-dichlorophenoxyacetic acid (2,4-D) (herbicide) and flumequine (bactericide) on synthetic goethite are presented here. The nature of interactions between adsorbate and adsorbent has been investigated by solution depletion, electrophoretic mobility, and infrared spectroscopic methods.

The adsorption has been found to be due to the interactions between the surface charge of goethite and the ionized acids as pH plays an important role. In the case of flumequine, which adsorbs more strongly than 2,4-D, the possibility of adsorption of neutral molecules also exists. While the results of all the methods complement each other in the case of flumequine, there seems to be some discrepancy in the case of 2,4-D. Solution depletion results indicate weak interactions whereas infrared and electrophoretic mobility measurements show the possibility of specific interactions for 2,4-D. The results are discussed to arrive at a possible mechanism of adsorption.

I. INTRODUCTION

Adsorption-desorption of toxicant agrochemicals on soil constituents is one of the major phenomena affecting their behavior in soil environment.[1] Biological

0-87371-616-7/93/$0.00 + $.50

efficacy of a pesticide may be increased (but only to a certain extent) by developing formulations suited to the specific needs.[2] However, little can be done to alter their behavior in soils. Thus, it is necessary to thoroughly examine their adsorption characteristics not only to obtain better agricultural yields, but also to avoid environmental hazards. Each method used for this purpose has its limitations, and conclusions drawn from the results of a particular method sometimes can be misleading. This aspect of interfacial studies has been well explained by Lyklema's[3] discussing the limitations of various methods. Zeltner et al.[4] also suggested using several methods to understand interfacial interactions.

In the present study, three methods (solution depletion, electrophoresis, and infrared (IR) spectroscopy) have been used to examine adsorption characteristics of two agrochemicals: 2,4-dichlorophenyloxyacetic acid (2,4-D) (herbicide) and flumequine (bactericide) on a synthetic goethite. 2,4-D is a well-known agrochemical used for its selective activity against broad-leaved weeds. Several studies have been reported[5-13] concerning its adsorption behavior on various soil constituents. Of these, a few[11-13] are devoted to the adsorption on goethite. Flumequine is a bactericide, generally used for human and veterinary therapeutics. In France, it has been registered recently for its agricultural use against fire blight (*Erwinia amylovora*). This has been reported[14] to adsorb strongly on soil (in Versailles, France) and does not desorb easily. Its desorption as a function of concentration from kaolinite was also found[15] to be difficult. The present work is an extension of the results presented earlier[16] on the adsorption of this chemical on goethite.

II. EXPERIMENT

A. Materials

The methods of preparation and identification of goethite used here have been described elsewhere.[17] The sample has a BET nitrogen surface area of 38 m^2/g. Its isoelectric point (IEP) as determined by microelectrophoresis is equal to 8.5. High purity 2,4-D and flumequine (structures showed below) are used, and their labeled ^{14}C as well as unlabeled products were procured from Amersham and Riker-3M, (France), respectively.

2,4-D, mol wt = 221.04, pKa = 2.75.

Flumequine, mol wt $= 261.25$, pKa $= 6.0$ (as determined by isobestic point method using UV spectrophotometer).

B. Methods

1. Adsorption Measurements

Adsorption studies were conducted by the solution depletion method using 2.5 g/dm^{-3} of goethite in all the cases. Goethite suspension (20 mL) mixed with the desired amount of adsorbate was stirred (at 20°C) for different time intervals (for kinetics) and centrifuged. The supernatant was analyzed for the adsorbate remaining unadsorbed using a liquid scintillation counter SL-4000 (Intertechnique, France).

For the effect of pH, goethite (2.5 g/dm^{-3}) was mixed with flumequine (80 μmol/dm^{-3} in 10^{-3} M KNO$_3$ medium) or 2,4-D (1800 μmol/dm^{-3} in 10^{-2} M KNO$_3$ medium), and analyzed for adsorption as stated above. Before mixing, each was adjusted at desired pH separately using HNO$_3$ and KOH.

The adsorption isotherms were obtained at pH static conditions using an automatic titrator (Urectron 6, Tacussel). Goethite suspension was mixed with different concentrations of flumequine (in 10^{-3} M KNO$_3$ medium) or 2,4-D (in 10^{-2} M KNO$_3$ medium) in a thermostated (20°C) cell. The quantity of chemical adsorbed was determined after 4 hr of mixing as described above.

2. Infrared Spectroscopic Measurements

IR spectra (transmission) were obtained on the IR spectrophotometer 580 (Perkin Elmer). 2,4-D and flumequine were analyzed using KBr pellets made by grinding 1 mg of the compound with 300 mg of KBr. Goethite with or without adsorbed acids was analyzed using self-supporting films made by spreading and drying (at ambient temperature) of 5 mL of the sample (5 mg goethite).

3. Electrophoretic Mobility Measurements

They were conducted using a modified[17] laser Zee Meter model 501 apparatus (Pen Kem, Inc.). The effect of 2,4-D and flumequine as a function of pH was seen on the electrophoretic mobility of goethite (4 g/dm^{-3} in 10^{-3} M KNO$_3$ medium). Measurements of pH were done before and after each determination of electrophoretic mobility.

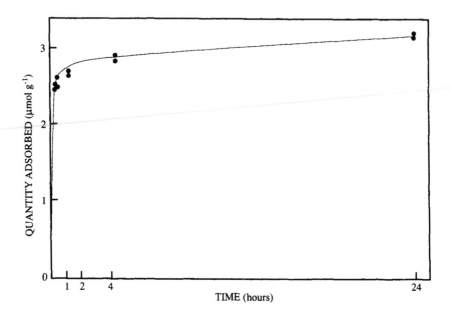

Figure 1. Adsorption of flumequine with time at pH 5 in 10^{-3} M KNO$_3$ medium.

III. RESULTS AND DISCUSSIONS

A. Adsorption Measurements

1. Kinetics

Adsorption equilibrium of 2,4-D (results are not shown) is reached within 8 hr. In the case of flumequine (Figure 1) adsorption is rapid initially and then continues slowly without reaching an equilibrium in the true sense. Similar results were reported[18,19] for adsorption of phosphate on goethite. The slower adsorption was suggested to be due to several processes, i.e., molecular rearrangement, induced precipitation on the surface, diffusion into the already adsorbed layers, and diffusion into micropores of the oxide. Because the goethite used here was found to be nonporous,[17] the possibility of diffusion into its pores is ruled out in the present case. Intermolecular attraction and/or diffusion into the already adsorbed layers appears to be the reasons for continued adsorption of flumequine with time.

2. Adsorption Isotherms

Adsorption isotherms (Figure 2a and b) at two pH values of 2,4-D are of the S type and of flumequine are of the L-type.[20] It shows that compared to 2,4-D,

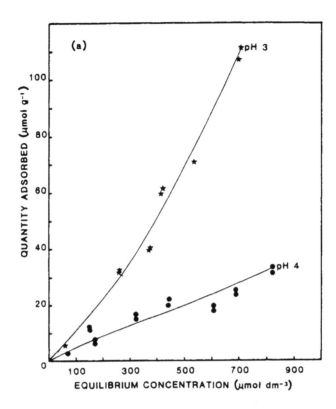

Figure 2. Adsorption of isotherms of 2,4-D (a) and flumequine (b) in KNO$_3$ medium.

flumequine adsorbs with stronger interactions. In the case of 2,4-D, it may be seen that increase in adsorption with concentration is more at lower pH 3. A phenomenon of flotation is also observed at the initial concentration of around 300 μmol/dm^{-3} for both pH values (3 and 4). Flotation results from the formation of a hydrophobic surface as the aromatic part of adsorbed 2,4-D remains directed in solution as reported earlier.[11]

3. Effect of pH

The results of adsorption as a function of pH (Figure 3) show that both the acids exhibit a maximum. As seen[21] in the case of monobasic acids, adsorption occurs at pH values lower than the IEP (pH 8.5) of goethite when the surface charge is positive. Generally, adsorption maxima occur at pH values close to the pK of the acid as observed here in the case of 2,4-D (pK 2.7 and maxima at pH 2.5). Reasons for the effect of pH on adsorption are explained in the review by Barrow.[21] In the case of flumequine (Figure 2b) maximum adsorption occurs at a pH (5.0) lower than its pKa (6.0). This indicates that the adsorption

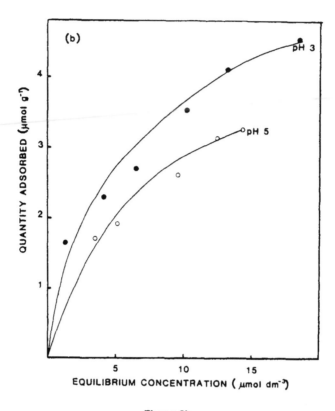

Figure 2b.

of neutral molecules is also possible. Adsorption at the pH beyond the IEP of goethite is not observed. This is, perhaps, due to the sharp increase in solubility of flumequine in water as indicated elsewhere[15] or due to the increase of the goethite negative charge.

Adsorption as a function of concentration and pH was found to be completely reversible in the case of 2,4-D, but partially reversible in the case of flumequine (results are not shown) indicating strong interactions of the latter with goethite surface.

B. Electrophoretic Measurements

Results of electrophoretic mobility with pH at various concentrations of acid adsorbates are shown in Figure 4. Both these acids bring a change in electrophoretic mobility of goethite and displace the isoelectric point. Such an effect was also shown by oxalate and phosphate ions while benzoate did not affect the mobility as well as the isoelectric point of the same goethite sample.[18] These

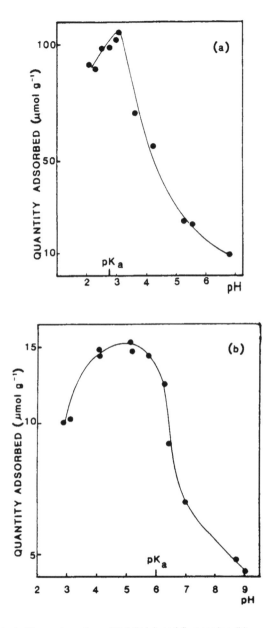

Figure 3. Effect of pH on adsorption of 2,4-D (a) and flumequine (b).

results with 2,4-D and flumequine show that both these acids adsorb on goethite by the forces that are stronger than Coulombic interactions, like oxalate and phosphate. The forces are stronger in the case of flumequine than in the case of

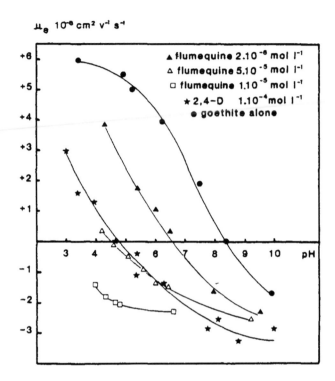

μ_e 10^{-8} cm² v⁻¹ s⁻¹

- ▲ flumequine 2.10⁻⁶ mol l⁻¹
- △ flumequine 5.10⁻⁵ mol l⁻¹
- □ flumequine 1.10⁻⁵ mol l⁻¹
- ★ 2,4-D   1.10⁻⁴ mol l⁻¹
- ● goethite alone

Figure 4. Electrophoretic mobility of goethite in function of pH in absence and in presence of acid adsorbates.

2,4-D. The only reported[12] study on electrophoretic mobility measurements on goethite in the presence of phenoxyacetic acids (including 2,4-D) indicated no displacement of isoelectric point (in NaCl medium), but a significant change in electrophoretic mobility especially at higher concentration. The results of the present study may be different from those reported, perhaps due to the difference in the nature of electrolytic medium used for electrophoretic measurements.

C. Infrared Spectroscopic Measurements

IR spectra of goethite alone and in the presence of anions are shown in Figure 5a and b. Goethite alone shows only two bands (3487 and 3660/cm) in the higher wave number region. Some of the reported studies presented three bands due to the OH group at three different sites, A, B, and C, of the goethite surface plane. The OH groups at sites A, B, and C were described[22] as coordinated to one, three, and two Fe^{3+} ions, respectively. The assignment of the bands in the higher wave number region is a matter of controversy.[23] Nevertheless, displacement in their positions can still be a useful tool to study any exchange of OH groups with adsorbed anions. The two bands observed at 1790 and 1662/cm are due to

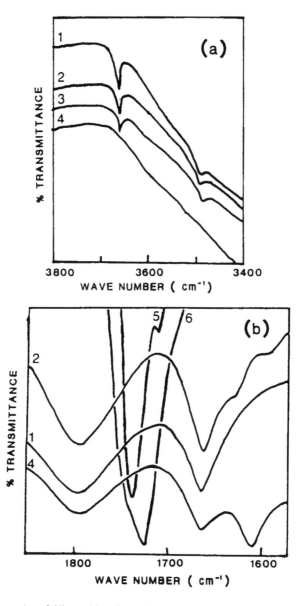

Figure 5. IR spectra of (1) goethite alone; (2) goethite with flumequine (12 μmol/g); (3) goethite with 2,4-D (25 μmol/g); (4) goethite with 2,4-D (45 μmol/g); (5) flumequine alone; and (6) 2,4-D alone.

the combination and overtone of the in-plane and out-of-plane deformations. These bands remain unperturbed in the presence of adsorbed species.

IR spectra of the sample adsorbed with 2,4-D show two visible changes: the disappearance of the two bands due to OH groups at A, B, and C sites and significant displacement of the band due to −C=O bond of crystalline 2,4-D toward the lower wave number. The disappearance of bands of the OH group clearly shows the exchange of these groups by the carboxylate form of 2,4-D. This is accompanied by the large displacement of band due to −C=O of COOH (1735/cm) appearing as that of the carboxylate (1608/cm) group.[24] The present study shows the possibility of complex formation facilitating the adsorption of 2,4-D by the exchange of all types of OH groups of goethite. In one of the earlier reported[7] studies on adsorption of 2,4-D on montmorillonite, it was shown that the adsorbed species are the molecular form of 2,4-D. Formation of the complex was found to occur only in the case of clay exchanged with a high polarizing cation such as Mg.

IR results of the flumequine adsorbed sample show similar changes except that the bands due to OH groups (high wave number) still persist. This is due to the fact that this was not feasible to take spectra at higher concentration of adsorbate due to its low solubility in water. The complex formation, however, can be visualized from the displacement of band due to −C=O bond of −COOH (1725/cm) to that due to −C=O of carboxylate (1637/cm).

IV. CONCLUSIONS

Results of different measurements conducted in the present study lead to the following conclusions:

1. Both acids adsorb due to interactions controlled by the pH, and hence the ionized part (−COO⁻) of the molecule plays an important role in their adsorption. Adsorption of the neutral molecule is also possible in the case of flumequine.
2. Results of all the measurements lead to the conclusion that flumequine adsorbs with strong interactions.
3. In the case of 2,4-D, the results of electrophoretic mobility and spectroscopic measurements show that 2,4-D also adsorbs due to specific interactions. However, the adsorption measurement indicated nonspecific interactions. This discrepancy is due to the limitations of these methods to measure adsorbed ions.

Because the spectroscopic measurements are conduced with dried samples, their results cannot be assumed to be similar to those of an aqueous medium (e.g., adsorption measurements). Such contradictory conclusions based on spectroscopic and adsorption measurements also were reported[25] earlier. On the other hand, the electrokinetic measurements give the changes in the electrokinetic charge density which represent the changes in the net charge density within the plane of shear. These changes include not only those arising from the anion adsorption, but also those arising from the rearrangement of potential determining ions and from coadsorbed counterions.[26] Thus, the results of adsorption isotherm and electrophoretic mobility can be assumed to be correlated always as seen[18] in the case of oxalate adsorption on goethite. Thus, it may be said that 2,4-D adsorbs due to interactions which are not as strong as can be visualized by the solution depletion method. Such interactions are not enough to resist desorption on washing perhaps because the solubility of 2,4-D in water dominates.

REFERENCES

1. Bailey, G.W. and J.L. White. *Res. Rev.,* 32:29–92, (1970).
2. Flanagan, J., in "Industrial Production and Formulation of Pesticides in Developing Countries," United Nations Publication, Vienna ID/75 Vol. 1, (1972) pp. 75–122.
3. Lyklema, J. *Adsorption from Solution at the Solid/Liquid Interface,* G.D. Parfitt and C.H. Rochester, Eds. (London: Academic Press Inc., 1983), pp. 223–246.
4. Zeltner, W.A., E.C. Yost, M.L. Machesky, M.I. Tejedor-Tejedor, and M.A. Anderson. *Geochemical Processes at Mineral Surfaces,* ACS Symposium Series 323, J.A. Davis and K.F. Hayes, Eds. (Washington, DC: American Chemical Society, 1986), 142–161.
5. Frissel, M.J. and G.H. Bolt. *Soil Sci.* 94:284–291 (1962).
6. Weber, J.B., P.W. Perry, and R.P. Upchurch. *Soil Sci. Soc. Am. Proc.* 29:678–688 (1965).
7. Dieguez-Carbonell, D. and C. Rodriguez Pascual. "Pesticides," *Environ. Qual. Saf.* (Suppl. 3):237–242 (1975).
8. Miller, R.W. and S.D. Faust. *Adv. Chem. Ser.* 14:121–134 (1972).
9. Haque, R. and R. Sexton. *J. Colloid Interface Sci.* 27(4):818–827 (1968).
10. Leopold, A.C., P. Van Schaik, and M. Neal. *Weeds* 8:48–54 (1960).
11. Watson, J.R., A.M. Posner, and J.P. Quirk. *J. Soil Sci.* 24(4):503–511 (1973).
12. Kavanagh, B.V., A.M. Posner, and J.P. Quirk. *J. Colloid Interface Sci.* 61(3):545–553 (1977).
13. Kavanagh, B.V., A.M. Posner, and J.P. Quirk. *J. Soil Sci.* 31:33–39 (1980).
14. Kerhoas, L. and J.C. Thoisy-Dur. *Methodological Aspects of the Study of Pesticides Behaviour in Soil,* P. Jamet, Ed. (Versailles, France: INRA, 1988), pp. 195–200.
15. Khandal, R.K., J.C. Thoisy-Dur, and M. Terce. *Geoderma* 50, 95–107 (1991).
16. Thoisy-Dur, J.C., M. Djafer, M. Terce, and A.M. Tabareau. *Methodological Aspects of the Study of Pesticides Behaviour in Soil,* P. Jamet, Ed. (Versailles, France: INRA, 1988), pp. 37–47.
17. Djafer, M., I. Lamy, and M. Terce. *Prog. Colloid Polym. Sci.* 79:150–54 (1989).
18. Djafer, M., R.K. Khandal, and M. Terce. *Colloid and Surfaces* 54, 209–218 (1991).
19. Anderson, M.A., M.I. Tejedor-Tejedor, and R.R. Stanforth. *Environ. Sci. Technol.* 19(7): 632–637 (1985).
20. Giles, C.H., T.H. MacEwan, S.N. Nakhwa, and D. Smith. *J. Chem. Soc.* 3:3973–3993 (1960).
21. Barrow, N.J. *Adv. Agron.* 38:183–230 (1985).
22. Russel, G.D., R.L. Parfitt, A.R. Fraser, and V.C. Farmer. *Nature* 248:220–21 (1974).

23. Buckland, A.D., C.H. Rochester, and S.A. Topham. *J. Chem. Soc. Faraday Trans. 1*, 76:302–313 (1980).
24. Cornell, R.M. and P.W. Schindler. *Colloid Polymer Sci.* 258:1171–1175 (1980).
25. Hingston, F.J., A.M. Posner, and J.P. Quirk. *J. Soil Sci.* 23:177–192 (1972).
26. Hough, D.B. and Rendall, H.M. in *Adsorption from Solution at the Solid/Liquid Interface*, (London: Academic Press Inc., 1983), pp. 247–319.

CHAPTER 7

Adsorption of Maleic Hydrazide on Mineral Surfaces

María C. Hermosín, Isabel Roldán, and Juan Cornejo

ABSTRACT

The adsorption mechanism of maleic hydrazide (MH) on diverse soil mineral surfaces has been elucidated by using pure minerals as model systems and carrying adsorption isotherms from aqueous solutions. The chosen adsorbents represented different mineral surface present in soils: (1) phyllosilicates as minerals of permanent charge (smectites: SAz, STx, and SWy) and (2) iron oxides (hematites, H) and oxihydroxides (lepidocrocite, L, and goethite, G) as minerals of variable charge. The shapes of isotherms showed MH adsorption on external surfaces for SAz, STx, L, G, and H samples whereas for SWy some penetration in the intracrystalline region is suggested. For SAz, STx, L, G, and H samples the adsorption of MH occurred on the external surface hydroxyls by polar bonds. For the smectite SWy due to its low layer charge, MH was adsorbed also at the interlamellar spaces probably associated to the interlayer cation by polar bonds.

I. INTRODUCTION

The organic and mineral components of the finest soil fraction (i.e., clay fraction) are the most active participants on the interface process of adsorption-

0-87371-616-7/93/$0.00 + $.50

Table 1. Surface Properties of the Minerals Used in This Study

Mineral	SAz	STx	SWy	H	L	G
C.C.C. meq/100 g	120.0	84.4	76.4	—	—	—
Se (BET) m²/g	97.4	83.8	31.8	17.3	116.0	80.0

desorption of organic chemicals in soils and in aquatic systems. The maleic hydrazide (MH) adsorption by soils has been related to their clay content,[1-3] and even more to the mineral composition of this clay fraction. However, it is difficult to determine the mechanism of the adsorption process at a molecular level in heterogeneous systems, like soils. For this reason it is very useful to study the interaction of an organic chemical with pure clay minerals as a simplified system, in order to add information on the behavior of a chemical in the soil-water systems.[4-6]

The objective of this research was to determine the adsorption mechanism of agrochemical MH on mineral components of the soil colloidal fraction or particulate matter suspended in surface waters. Different minerals were selected as model adsorbents: three montmorillonite (SAz, STx, and SWy) as phyllosilicates of permanent charge and three iron oxides (lepidocrocite, L; goethite, G; and hematite, H) as models of variable charge surfaces.

II. MATERIALS AND METHODS

The maleic hydrazide (6-hydroxy-2H-pyridazine-3-one) used was purchased from Fluka as a pure compound. MH is a growth regulator and inhibitor, sometimes used as an herbicide by itself or combined with some other compounds. The minerals used as model adsorbents were chosen as representing a diversity of structural and surface properties covering the variability of mineral surfaces present in soils, sediments, or particulate matter suspended in natural waters: (1) three montmorillonites (SAz, STx, and SWy[7]) representing the phyllosilicates with a permanent negative charge due to isomorphic substitutions (Mg^{2+} for Al^{3+}) in their structure, which is compensated by inorganic cations that can be easily interchanged; and (2) three iron oxides (lepidocrocite, goethite, and hematites) as representative surfaces of variable charge due to their hydroxilated (Fe-OH) surfaces. The surface properties of these minerals are summarized on Table 1.

The adsorption isotherms were carried out by weighing duplicate 0.2 g of mineral in 50 mL polyethylene centrifuge tubes with screw caps and adding 20 mL of 0 (blank)-3 mM MH aqueous solutions. The suspensions were shaken for 24 hr. Previously this time had been verified to be sufficient for reaching the adsorption equilibrium. After that the tubes were centrifuged to 15,000 rpm, and the supernatants were analyzed by UV spectroscopy at 302 nm. The differences between the initial and final concentrations were assumed to be adsorbed.

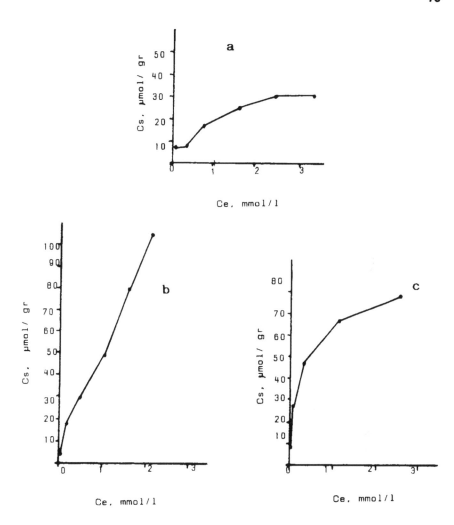

Figure 1. Adsorption isotherms of MH in (a) SAz, (b) SWy, and (c) lepidocrocite.

The adsorption isotherms were obtained by plotting the amount of MH adsorbed, Cs (μmol/g), vs the equilibrium concentration, Ce (mmol/L).[2,8]

III. RESULTS AND DISCUSSION

The adsorption isotherms are shown in Figure 1. All adsorption isotherms are of the L type according to the Giles et al.[9] classification, although some of them (such as SWy) showed a certain C character. It is observed that there is a great difference among the montmorillonites: SAz and STx showed lower adsorption than SWy. The oxides showed a high adsorption level for goethite and lepidocrocite and a lower level for hematites. The shapes of isotherms of MH on the diverse mineral surfaces studied are composed of an initial convex curvature indicating an initial similar behavior; however, these shapes changed (as equilibrium concentration increased) to an L type for SAz, STx, H, L, and G; and a C type for SWy.

According to the Giles et al.,[9] the L isotherm indicated a flat adsorption of polar molecules on a polar substrate with adsorption sites are limited to a monolayer that is reached at the plateau, as shown for MH adsorption on SAz, STx, H, L, and G. The C type isotherms[9] indicated a constant partition between substrate and solution, i.e., as more solute is adsorbed more sites must be created. This could indicate some penetration of MH in the interlamellar surfaces of the SWy.

The adsorption data were assayed to fit the Langmuir and Freundlich equations. The Langmuir equation can be written as follows:

$$Cs = Cm \frac{CeL}{1 + CeL} \tag{1}$$

where
Cs = the amount adsorbed
Ce = concentration of solute at equilibrium
L and Cm = characteristics of the systems

Cm is considered the amount adsorbed for a monolayer covering the sorbent surface which is reached when the isotherms show a plateau, and thus Cm is a measure of the adsorption capacity.

The Freundlich equation is generally expressed as follows:

$$Cs = K_f Ce^n \tag{2}$$

where
Cs = amount adsorbed (μmol/g)
Ce = equilibrium concentration of the adsorbate (mmol/L)
Kf and n = characteristic paramaters of the adsorbent-adsorbate system

Kf is known as adsorption capacity, because it deals with Cs for Ce = 1 and n is the intensity factor of the adsorption. These parameters are summarized on Table 2 for the adsorbents studied. The adsorption capacity of these minerals showed a different variation when measured by Cm (SWy, G, L, H, STx, SAz) or K_f (G, L, SWy, H, STx, SAz) because K_f is calculated in the range of concentration of MH used in the experimental studies, whereas Cm is calculated by extrapolation of the experimental data, especially for SWy with an isotherm that did not reach a plateau. On Table 2 the surface coverage for MH (Cm/Se or Kf/Se) of the diverse minerals is summarized. As can be seen, whatever the

Table 2. Characteristics and MH Surface Covering of Minerals Studied

Minerals	Cm (μmol)	Cm/Se (μmol/m²)	K$_f$ (μmol/g)	K$_f$/Se (μmol/m²)
SAz-1	38.8	0.39	18.1	0.18
STx-1	31.3	0.37	19.6	0.23
SWy-1	206.1	6.48	54.0	1.70
Goethite	166.0	2.07	86.5	1.08
Lepidocrocite	77.4	0.67	59.7	0.52
Hematite	47.1	2.72	24.8	1.43

capacity parameter considered (Cm or Kf) the value of surface coverage for SWy is much higher than those corresponding to others adsorbents.

Taking into account the nature of the mineral surfaces studied.[10,11] there are two types of active adsorption centers to be considered for polar molecules such as MH:hydroxyl surface groups and exchangeable cations. Oxides and oxyhydroxides have an important number of surface hydroxyl groups and no exchangeable cations whereas the phyllosilicates contain few hydroxyl groups, but the exchangeable cations in the interlamellar surfaces are very active centers for adsorption.

The adsorption capacity and surface coverage values seem to indicate that the MH adsorption on Fe hydrous oxide and oxide, and SAz and STx montmorillonites should occur on surface hydroxyls because the sequence of a number of these groups was similar to that of the adsorption capacity values. However, SWy, also having a low number of surface hydroxyls, presented much higher MH adsorption capacity and surface coverage than SAz and STx; this suggests that in this montmorillonite, MH should be adsorbed on internal surfaces. MH molecules could be bound by polar bonds to the exchangeable cations, as it has been shown for others polar molecules.[4,12,13]

The reason for MH adsorption in the interlamellar spaces of SWy (whereas it is only adsorbed on the external surface of SAz and STx) was the higher charge of these last minerals. Effectively the high charge of SAz and STx (both having an interlayer cation of calcium) makes the interaction layer-cation-layer stronger than in the case of SWy. Thus this spacing was less accessible for the adsorption of MH.

To check the interlamellar adsorption on SWy, the adsorption of MH was carried out on this sample previously saturated with the inorganic cations: Na, K, Ca, and Fe. The adsorption isotherms are shown on Figure 2, and the corresponding Langmuir and Freundlich parameters are summarized on Table 3. As was expected, MH adsorption isotherms were affected by the interlayer cation which determines the shapes of isotherms that were of the C type for Ca SWy and K SWy, L type for Na SWy, and S type for Fe SWy. This is a consequence of the accessibility degree of interlamellar spacing determined by the interaction layer-cation-layer. Effectively the adsorption capacity was inversely related to

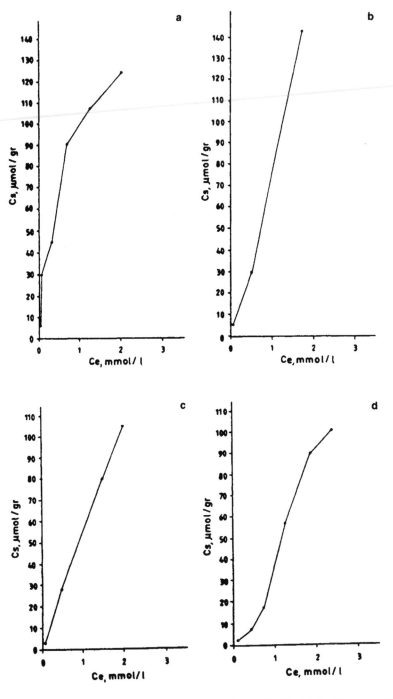

Figure 2. Adsorption isotherms of MH in (a) Na SWy, (b) K SWy, (c) Ca SWy, and (d) Fe SWy.

Table 3. Adsorption Parameters and Correlation Coefficients
Obtained from Langmuir and Freundlich Equations
for Experimental Data

Minerals	Cm (μmol/g)	L	r	K_f (μmol/g)	n	r
Na SWy	134.7	4.99	0.96	96.7	0.60	0.95
K SWy	745.5	0.12	0.21	70.2	0.83	0.98
Ca SWy	—	—	—	52.7	1.01	0.99
Fe SWy	—	—	—	32.8	1.34	0.99

the charge of the cations and their ionic potential, indicating an association of
MH to the interlayer cation by polar bonds.[12,14]

Taking into account the polar character of MH molecules and the surfaces of
the minerals studied, MH adsorption can be schematized as follows:

(1)   For montmorillonites:

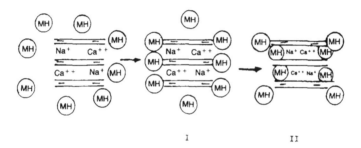

Process I, bonding MH by polar bonds to hydroxyl surface groups, occurred
for SAz, STx, and SWy. On the other hand, process II occurred only on SWy
due to its lower layer charge[7] that makes the layer opening easier, thus MH
would be associated to the exchangeable cation by ion-dipole bonds.

(2)   For iron oxides surfaces with a high density of OH groups:

This adsorption is the same as that occurring in SAz and STx by polar bonds
probably between surfaces OH and carbonyl groups of MH. The higher adsorption capacities of L and G are due to the higher number of OH in their surfaces
as compared to SAz, STx, and H.

IV. CONCLUSIONS

The adsorption of MH occurs through different mechanisms on the adsorbent studied:

1.   On montmorillonite, MH is adsorbed on external surfaces; and when the layer charge of this mineral is low, it is on the interlamellar spaces as a polar molecule associated to the interlayer cation.
2.   On iron hydrous oxide and oxide the adsorption of MH occurred on external surfaces by H bonds probably between the C=O groups of MH and surface hydroxyls.

ACKNOWLEDGMENT

This work has been supported by funds of the Junta de Andalucía and the DGICYT (Project No. NAT90-910).

REFERENCES

1. Hermosín, M.C., J.L. Perez-Rodriguez, and J. Cornejo. "Adsorption-Desorption of Maleic Hydrazide as a Function of the Soil Properties," *Soil Sci.* 144:250–257 (1987).
2. Hermosín, M.C. and J. Cornejo. "Maleic Hydrazide Adsorption by Soils and the Use of the Mole Fraction Form of the Freundlich Equation," *Soil Sci.* 144:453–456 (1987).
3. Hermosín, M.C. and J. Cornejo. "Assessing Factors Related to Pesticide Adsorption by Soils," *Toxicol. Environ. Chem.* 25:45–55 (1989).
4. Aochi, Y. and W.J. Farmer. "Infrared Spectroscopy of Picloram Interaction with Al(III)-, Fe(III)- and Cu(II)-Saturated and Hydrous Oxide-Coated Montmorillonite," *Clays Clay Miner.* 29:191–197 (1981).
5. Ainsworth, C.C., J.M. Zachara, and R.L. Schmidt. "Quinoline Sorption on Na-Montmorillonite: Contributions on the Protonated and Neutral Species," *Clays Clay Miner.* 35:121–128 (1987).
6. Ferreiro, E.A., S.G. Bussetti, and A.K. Helmy. "Sorption of 8-Hydroxy-Quinoline by Some Clays and Oxides," *Clays Clay Miner.* 36:61–67 (1988).
7. Van Olphen, H. and J. Fripiat. *Data Handbook for Clay Materials and Other Non-Metallic Minerals* (New York: Pergamon Press, 1979), 346 pp.
8. Bowman, B.T. "Anomalies in the Log Freundlich Equation Resulting in Deviations in Adsorption K Values of Pesticides and Other Organic Compounds When the System of Units is Changed," *J. Environ. Sci. Health* B16:113–123 (1981).
9. Giles, C.H., T.H. MacEwan, S.N. Nakhwa, and D. Smith. "Studies in Adsorption. Part XI. A System of Classification of Solution Adsorption Isotherms and Its Use in Diagnosis of Adsorption Mechanisms and in Measurements of Specific Surface Area." *J. Chem. Soc.* 3973–3993 (1960).

10. Sposito, G. "The Reactive Solid Surfaces in Soils," in *The Surface Chemistry of Soils* (New York: Oxford University Press, 1984), pp. 1–46.
11. Greenland, D.J. and C.J.B. Mott. "Surfaces of Soil Particles," in *The Chemistry of Soils Constituents,* D.J. Greenland and H.B. Hayes, Eds. (New York: John Wiley & Sons, 1978), pp. 321–352.
12. Mortland, M.M. "Clay-Organic Complexes and Interactions," *Adv. Agron.* 22:75–115 (1970).
13. Sanchez-Martin, M.J. and M. Sanchez-Camazano. "Adsorption of Chloridazone by Montmorillonite," *Chemosphere* 16:937–944 (1987).
14. Sposito, G. "Inorganic and Organic Solute Adsorption in Soils," in *The Surface Chemistry of Soils* (New York: Oxford University Press, 1984), pp. 113–153.

CHAPTER 8

The Euro-Soil Concept as a Basis for Chemicals Testing and Pesticide Research

Gerald Kuhnt

I. INTRODUCTION

Today, an immense number of investigations are conducted on the behavior and fate of pesticides in the environment. On the basis of all the detailed information available it should be possible to make well-founded hazard predictions and risk assessments. In most cases, however, there are numerous problems in combining all the data, using them for modeling and extrapolation purposes, and giving detailed recommendations for the control and limitation of pesticide use. The main reason for this is that most of the research results are not directly comparable and are only valid for small areas or specific situations. Therefore, a system has been developed to identify soil types being representative for larger areas. If samples from these soils (reflecting a wide variety of sorption controlling properties) are used for chemicals testing, the results can form a basis for realistic hazard assessments.[1]

II. BASIC PHILOSOPHY OF THE EURO-SOIL PROJECT

In 1985, the German Federal Environmental Agency and the Commission of European Communities launched the Euro-Soil Project on the basis of the following situation.

0-87371-616-7/93/$0.00 + $.50

In connection with the intended harmonization of testing procedures for new chemicals in the European Community (EC), soils had to be selected for use in adsorption/desorption testing because in the future a new chemical, once tested and regarded to be nonhazardous in one EC member state, is allowed to be distributed throughout the Community without further testing.[2]

Therefore, the method for identification of representative soils should be chosen based on the fact that the limited number of soils to be used for testing ought to represent both a maximum area of the European Community and the wide variability of relevant parameters responsible for sorption processes in soils.[3]

In detail, three basic requirements have to be fulfilled:

1. Representativity of soil characteristics — The selected soils must cover a wide range of sorption-controlling properties so that basic parameters of the potential behavior of a certain chemical in soil can be assessed by comparing various test results.
2. Representativity with respect to frequency distribution — If the reference soils are typical representatives of the soil associations that are most widespread within the EC, each test result is to a certain extent valid for a large area of the Community.
3. Regional representativity — Considering that in most cases the association patterns of different soil types reflect important factors of soil formation and development (such as geology, geomorphology, topography, or climate[4]), the spatial distribution of the various soil units must be defined. Therefore, the soils sampled at the selected locations should be associated with other pedological units in such a manner that they are regionally representative of the EC member states.

Since climatic conditions have both a direct and an indirect influence on important pedological factors, the selected sampling sites should be appropriately distributed to adequately reflect the main climatic zones of the EC territory.

Soils and vegetational cover interact in manifold ways; therefore the main types of plant communities and of land use patterns in the European Community must be taken into account similarly. In compliance with the requirements mentioned above, the selection of soil samples is appropriately based on a five-level approach.

The first step is the evaluation of small-scale maps in order to define the typical and most frequent soils of the European Community. For this purpose the "Soil Map of the European Communities 1:1 Mio." was digitized, because only maps yield the area-covering information necessary to calculate frequency distributions. Since soil properties are also dependent on other factors such as vegetation, land use, or climate, the respective data were collected and converted in a way that they can form additional levels of information. This implies the possibility of more precisely defining the environmental boundary conditions under which certain soil types most frequently occur.

The second step is the nearest-neighborhood analysis which leads to determination of regionally representative sampling sites. This geostatistical procedure is able to unfold the most typical neighborhood relationships and association patterns of the soil types as they appear on the map. From various investigations it was found that most of the relationships are genetically determined. Due to this reason the neighborhood analysis is a capable tool for the identification of regionally representative sampling sites.[5]

In a third step a comprehensive study of literature and the evaluation of large-scale maps as well as metric soil profile data are necessary to ascertain whether the defined representative soils adequately reflect wide variability of the whole soil inventory from an ecochemical point of view. This means that the soils selected have to cover the range of sorption-controlling properties mainly occurring in the soils of the European Community.

Step four consists of the verification of theoretical investigations by visual inspection in the field, including site exploration and geological and pedological mapping, to finally locate discrete soil profiles where samples are taken and analyzed to determine the validity of the selection.

Step five is the treatment, homogenization, and bottling of the soil samples in order to store them in a soil bank. This makes reference material available which can be used as a European standard for the identification of similar soils in various EC Member States.

III. REPRESENTATIVE SOILS OF THE EUROPEAN COMMUNITY

By combining the total of all information gained, five EC reference soils and the corresponding sampling locations were identified (cf. Table 1):

Euro-Soil 1 — Vertic cambisol	Sicily	Italy
Euro-Soil 2 — Rendzina	Peloponnisos	Greece
Euro-Soil 3 — Dystric cambisol	Wales	United Kingdom
Euro-Soil 4 — Orthic luvisol	Normandy	France
Euro-Soil 5 — Orthic podzol	Schleswig-Holstein	Germany

Cambisols, luvisols, podzols, and rendzinas are the most frequent soils of the European Commmunity, covering about 70% of the area. The dominant soil moisture regimes within the EC realm are udic and xeric, and the dominant temperature regimes are mesic and thermic. Cambisols are distributed throughout the Community; therefore, two representatives of this group under different climatic conditions are considered. Luvisols mainly occur in North and Central Europe, Rendzinas are frequent in the South, and podzols are typical soils of the North. Accordingly, the sampling sites are located within the climatic zones where these soils predominantly occur.

Table 1. EC Representative Soils for Chemicals Testing

EC soil map 1:1 Mio.	Cambisols		Luvisols	Podzols	Rendzinas
FAO soil map of Europe	Brown forest soils p.p.	Brown Mediterranean soils p.p.	Gray-brown podzolic soils	Podzolized soils	Rendzinas
Frequency (%)	44.7		15.7	6.7	5.0
Soil climate Moisture regime Temperature regime	Udic Mesic	Xeric Thermic	Udic Mesic	Udic Mesic	Xeric Thermic
Vegetation/Land use	Pasture	Meadow	Arable ground	Coniferous forest	Broad-leaved trees/scrub
Geology/parent material	Till Glacial drift	Marine Deposits	Loess	Fluvioglacial sediments	Lacustrine deposits
FAO soil unit	Dystric cambisol	Vertic cambisol	Orthic luvisol	Orthic podzol	Orthic rendzina
Representative sampling	Radyr	Aliminusa	Rots	Gudow	Souli
Location	Wales	Sicily	Normandy	Schleswig-Holstein	Peloponnisos
EC member state	United Kingdom	Italy	France	Germany	Greece

From a computerized analysis of maps of the natural vegetation and land use in the European Community it was found that pasture and meadow, arable ground, coniferous forest, and broad-leaved trees must be taken into account to assure representativity. As a consequence, during field work special attention was also paid to these requirements.

The synopsis of the geological situation demonstrates that a reasonable diversity in the parent material of soil formation was also achieved. The combination of different soil types on alternate parent material under varying climatic conditions and numerous types of vegetation forms the best preconditions for obtained reference material that either is representative of the EC territory or differs with respect to its sorptive properties.

After a comprehensive evaluation and documentation of the representative locations and profiles (cf. Figure 1), ca. 150 kg of topsoil material was sampled from the A horizons, closely adjacent to the profile pit. Before that, numerous drill cores in the neighborhood of the profile were taken to ensure that changes in soil quality and structure do not occur and that the profiles are characteristic of the sampling area as a whole. In addition, one subsoil — the B horizon of the orthic luvisol — has been sampled to study the chemical behavior in deeper layers of the profile. Before taking the specimens, litter and plants had to be removed carefully to guarantee proper sampling and maximum homogeneity. The specimens were immediately transported to the EEC Joint Research Center (JRC) Ispra/Italy for air-drying and further treatment.

To obtain air-dried fine soil, the material had been dried in air-conditioned labs for 3 to 4 months. Before sieving, the aggregates were crushed by hand because using mechanical hammers or ball mills influences the test results in an unpredictable way. The sieved soil material was then transferred into large homogenization drums and after 1 week of continuous rotation, put into 500 g bottles. To prevent microbial degradation during tests, the material was sterilized by radiation. One part of the samples served for performing an intercomparison test on adsorption/desorption;[6] the remainder is stored under dry and cool conditions at the EEC JRC in Ispra/Italy.

IV. PROPERTIES OF THE EURO-SOILS

Table 2 shows the main pedological parameters of the reference soils. In general, it can be emphasized that nearly all of the parameters measured vary over wide ranges.

With respect to grain size distribution (cf. Figure 2) the vertic cambisol from Italy shows the typical dominance of clay while the podzol has a very high sand content. Due to the downward translocation of clay minerals, in the A horizon of orthic luvisols silt is enriched. The soils from Greece and the United Kingdom show more or less leveled grain size distributions.

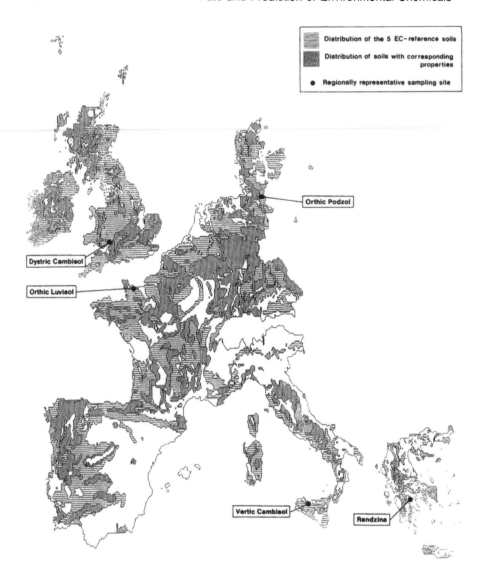

Figure 1. Distribution of EC representative soils and sampling sites.

Regarding the pH values (cf. Figure 3) a wide range is covered. While the podzol sampled in coniferous forests shows the expected acidity, the rendzina from Greece has pH values above neutrality due to the high amount of free calcium carbonate in the A horizon (cf. Table 2).

Also the organic matter content (Figure 4) varies within a wide range. As expected, the subsoil contains very little organic substance while the podzol,

Table 2. Main Pedological parameters of Euro-Soils

Pedological characterization of reference soil samples	Euro-Soil 1	Euro-Soil 2	Euro-Soil 3	Euro-Soil 4	Euro-Soil 5	Euro-Soil 6
Sand (total)	3.3	13.4	46.4	4.1	81.6	1.7
Coarse + medium %	2.0	4.4	23.1	1.1	64.8	0.3
Fine %	1.3	9.0	23.3	3.0	16.8	1.4
Silt (total)	21.9	64.1	36.8	75.7	12.7	82.4
Coarse %	4.0	21.3	19.4	52.2	7.4	62.5
Medium %	9.7	23.1	11.6	19.4	4.3	17.3
Fine %	8.2	19.7	5.8	4.1	1.0	2.6
Clay %	75.0	22.6	17.0	20.3	6.0	16.0
pH values						
Water	5.9	8.0	5.8	7.0	4.6	8.3
Calcium chloride	5.1	7.4	5.2	6.5	3.2	7.2
Potassium chloride	5.1	7.5	5.2	6.5	3.4	7.1
Total carbon %	1.5	10.9	3.7	1.7	10.9	0.3
$CaCO_3$ %	0.0	60.45	0.0	0.0	0.0	0.0
Organic carbon %	1.30	3.70	3.45	1.55	9.23	0.25
Organic matter %	2.65	6.4	6.44	2.86	15.92	0.78
N %	0.17	0.20	0.26	0.16	0.30	0.02
C/N ratio	7.65	18.50	13.27	9.69	30.77	12.5
Organic sulfur %	0.054	0.028	0.055	0.034	0.078	0.012
Total P %	0.15	0.15	0.38	0.29	0.21	0.15
CEC mval/100 g	29.9	28.3	18.3	17.5	32.7	11.4
Total Fe ppm	37050.0	9850.0	14370.0	11500.0	1040.0	12440.0
Amorphous Fe ‰	3.22	0.18	4.75	1.93	0.56	0.73
HCl soluble Fe ‰	1.820	0.002	2.200	1.470	0.105	1.140
Amorphous Al ‰	0.64	0.17	1.58	0.81	0.97	0.56
HCl soluble Al ‰	0.83	tr.	1.67	1.55	0.93	1.56
SiO_2 %	56.22	21.60	68.45	68.63	71.57	68.56

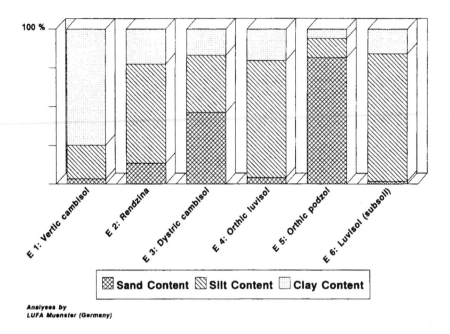

Figure 2. Grain size distribution of Euro-Soils.

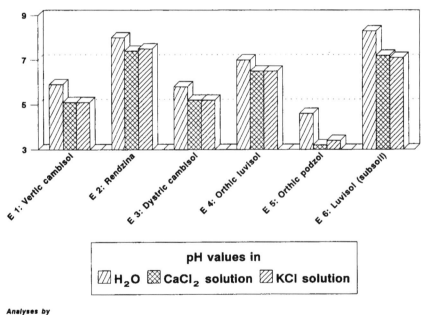

Figure 3. pH values of Euro-Soils.

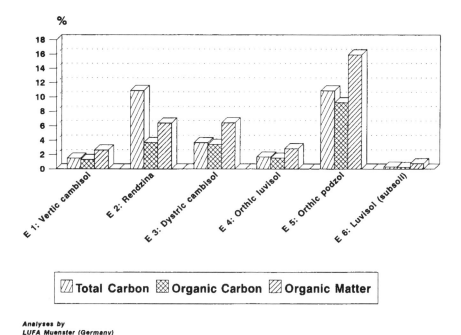

Figure 4. Organic matter content of Euro-Soils.

due to the inhibited microbial activity of this acid soil, contains high amounts of poorly humified organic matter. As to the completely different sorption controlling properties of Euro-Soils, it has to be expected that the sorption capacity also is dissimilar. Figure 5 shows some results of batch experiments, expressed as K' values. Lindane, for example, is adsorbed best by those soils showing average pH values and a higher clay or fine silt content. Organic matter seems to be of minor importance, because the vertic cambisol and the orthic luvisol contain less organic carbon than the other soils.

The tendency of the anionic substance 2,4-dichlorophenoxyacetic acid (2,4-D) to be adsorbed, particularly to humic substances at low pH values, is also clearly visible. Apart from the podzol, all soils (especially those with high pH values) show only poor sorption capacities for 2,4-D.

V. THE EURO-SOIL CONCEPT AS A BASIS FOR PESTICIDE RESEARCH

The Euro-Soil concept offers four main advantages which could be profitable for the implementation of a multinational and interdisciplinary program on pesticide research.

K' Value

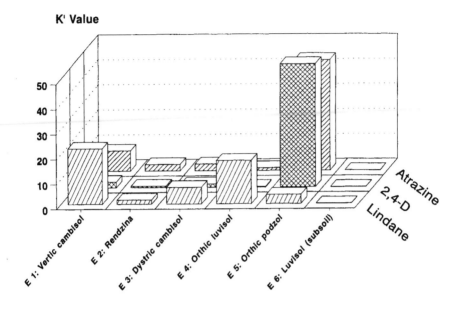

Mean Values
EC Intercomparison Test

Figure 5. Sorption capacities of Euro-Soils for different chemicals.

1. A method is available for the identification of representative soils and sampling sites. However, the Euro-Soils taken so far probably are not the best basis for pesticide research, because they have to be valid for all kinds of chemicals and the main conditions occurring in EC Europe. With respect to pesticide research, presumably some other soils have to be selected to emphasize the importance of agriculturally used soils. On the basis of the data collected and the methods developed, it is no problem at all to identify representative soils from a set of data being reduced to agricultural soils.

2. The application of the five-level approach consequently leads to the definition of soils with completely different properties. Therefore, a comparison of research results can form a sufficient basis for modeling and hazard assessments.

3. The material of the reference soils can always be taken in the same way and by the same persons, which is also very important. Furthermore, the material can be treated in the same manner and stored under appropriate conditions in order to have a standard available at all times.

4. Comprehensive analyses of the soils taken into consideration can be performed by one central laboratory and additional specialists so that a maximum amount of fully comparable data is available for correlation analyses and interpretation purposes.

Soils are highly complex three-phase entities, varying in space and time; and therefore only rough estimations are possible with respect to the behavior of pesticides in the numerous soil types existing. However, the Euro-Soil concept allows the knowledge and experience of various researchers, laboratories, and institutions to be brought together to produce comparable results which are valid for larger regions of the European Community.

ACKNOWLEDGMENTS

First of all I should like to thank Mr. Poremski and Dr. Herrmann of the German Federal Environmental Agency (UBA) for promoting the Euro-Soil Project. Furthermore, thanks are extended to the Commission of the European Communities for additional support, and especially to Dr. Murphy of DG XI for accompanying the work with professional zeal and outstanding commitment. I also want to thank Dr. Vetter, especially for his help during the development and adaptation of the geostatistical procedures. Moreover, I am much obliged to Professor Fierotti and Dr. Dazzi (Palermo), Professor Lautridou and Dr. Pellerin (Caen), Professor Yassoglou and Dr. Kosmas (Athens) as well as Dr. Janetzko (Kiel) and Mr. Richards (Radyr) for kindly lending their valuable aid and support in the field during the sampling campaigns. I am also grateful to Dr. H. Muntau and co-workers at the JRC Ispra/Italy for undertaking the treatment and homogenization of the soil material. Last, but not least, I thank the head of our working group, Professor Fränzle, most cordially for his kind readiness to accompany every step of the project.

REFERENCES

1. Fränzle, O. "Regionally Representative Sampling," in *Environmental Specimen Banking and Monitoring as Related to Banking*, R.A. Lewis, N. Stein, and C.W. Lewis, Eds. (Boston: Martinus Nijhoff Publishers, 1984), pp. 164–179.
2. Fränzle, O., G. Kuhnt, and L. Vetter. "Auswahl repräsentativer Böden im EG-Bereich," in G. Brümmer, et al., Eds. Fortschreibung der OECD-Prüfrichtlinie 'Adsorption/Desorption' im Hinblick auf die Übernahme in Anhang V der EG Richtlinie 79/831:Auswahl repräsentativer Böden im EG- Bereich und Abstufung der Testkonzeption nach Aussagekraft und Kosten. Umweltforschungsplan des Bundesministers für Umwelt, Naturschutz und Reaktorsicherheit, Forschungsbericht 106 02 045, Berlin (1987).
3. Kuhnt, G., O. Fränzle, and L. Vetter. "Regional repräsentative Auswahl von Böden für die Umweltprobenbank der Bundesrepublik Deutschland," *Kiel. Geogr. Schr.* 64:79–108 (1986).
4. Kuhnt, G. "Die grossräumige Vergesellschaftung von Böden. Rechnergestützte Erfassung pedogenetischer Zusammenhänge, dargestellt am Beispiel der Bundesrepublik Deutschland," *Erdkunde* 43:170–179 (1989).
5. Kuhnt, G. and L. Vetter. "Rechnergestützte Auswertung geowissenschaftlicher Karten als Grundlage der Umweltplanung," *Kartogr. Nachr.* 5:190–198 (1988).
6. Herrmann, M. "Evaluation of the EEC Laboratory Ringtest 'Absorption/Desorption of Chemicals in Soil' " in *Euro-Soils, Identification, Collection, Treatment, Characterization*, A. Kuhnt and H. Muntan, Eds. (JRC Ispra: 1992), pp. 78–142.

CHAPTER 9

Effect of Polymers on Adsorption of Flumequine on Kaolinite

R. K. Khandal, J. C. Dur, and M. Terce

ABSTRACT

Adsorption characteristics of flumequine on kaolinitic clay have been discussed. Effect of polymers (CMC, PVA, and polysaccharides) which are generally used as formulation adjuvants or as soil conditioners on the adsorption of flumequine has been examined in the present study. It has been observed that flumequine adsorbs strongly, and its desorption is difficult. pH played an important role in determining adsorption. In the presence of polymers, the adsorption exhibits three patterns: decrease, no change, and initial decrease followed by increase in the quantity adsorbed. Most of the ionic polymers have resulted in the decrease of adsorption due to the competition or flocculation effect. Nonionic polymers except PVA and scleroglucan showed little effect. In most of the cases the change in adsorption is small compared to the total quantity of flumequine adsorbed. This further shows that flumequine adsorbs with strong interaction which remain unaffected in the presence of polymers known for their strong affinity toward clay.

I. INTRODUCTION

Flumequine (a fluoroquinolone) is a bactericide generally used for human and veterinary therapeutics. Recently, it has been registered in France for its possible agricultural application against the bacteria (*Erwinia amylovora*) which causes the so-called "fire blight" disease in plants. It has been reported[1] that this bactericide adsorbs strongly on soil (of Versailles, France) and does not desorb easily. The nondegradability by microorganisms in soil and resistance to leaching of flumequine was attributed to its high adsorption by soil constituents. It has also been reported[2] to adsorb specifically on synthetic geothite. On the kaolinitic clay, its adsorption has been reported[3] to be unaffected even by the presence of phosphate ions which are known for their specific adsorption on kaolinite. In the present study, the effect of polymers (carboxymethyl cellulose; polyvinylalcohol; polygalacturonic acid; guar derivatives, i.e., HP_3, CP_{14}, T_{4246}; levan; and scleroglucan) on adsorption of flumequine on kaolinitic clay have been examined. These polymers have been chosen because of their industrial use as formulation adjuvants or as soil conditioners.

II. EXPERIMENTAL

A. Materials

Natural kaolinitc clay (St. Austell, UK) without further treatment has been used as an adsorbent. This clay has a surface area of 14 m^2/g and a cation exchange capacity of 12 meq/kg. Flumequine (structure shown below) of high purity (99.85%) has been used as the adsorbate. Its ^{14}C-labeled form has been procured from the 3 M-Riker Laboratory (Pithiviers, France).

Flumequine, mol wt = 261.25, pKa = 6.0.

Polymers used include polyvinyl alcohol (PVA), M_r 28,000, containing 12% acetate (Harlow Chemical Co. Ltd., UK); and carboxymethyl cellulose (CMC), hydrazide soluble in water (Sigma Chemicals, U.S.). Both polymers are nonionic. HP_8 (nonionic), T_{4246} (anionic), and CP_{14} (cationic) were the guar derivatives used where nonionic cis hydroxyl groups were substituted by hydroxypropyl, carboxyl, and quaternary ammonium groups, respectively.[4] These are the polysaccharides consisting of mannose and galactose units (Celanese Corporation, Louisville, KY, U.S.). Scleroglucan (nonionic) was the fungal slime from scleratium glycanium (CECA, France) of M_r 1.5×10^6. Levan (nonionic) was the product from *Acrobacter levanicum* (Sigma Chemicals); xanthan (anionic) was the product of *Xanthomonas campestris* (Rhône-Poulenc, France).

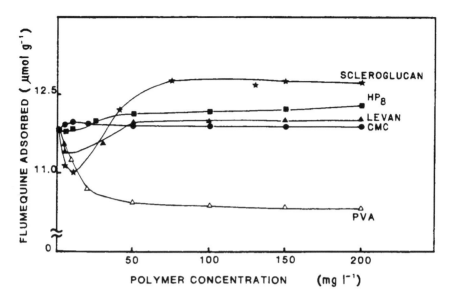

Figure 1. Effect of nonionic polymers on adsorption of flumequine.

B. Methods

Adsorption of flumequine in the presence of polymers at different concentrations was studied by use of the solution depletion method as described elsewhere.[3]

III. RESULTS AND DISCUSSIONS

A. Adsorption Behavior of Flumequine Alone

The results (not shown here) of adsorption-desorption of flumequine as a function of pH and concentration showed[3] that this bactericide adsorbs strongly on kaolinitic clay. The desorption is difficult in acidic conditions (pH < 6.0). Intermolecular interactions, solubility in water, and pH each played an important role in adsorption. It was observed that initial rapid adsorption was followed by slower but continued adsorption; equilibrium, in the true sense, did not appear to be reached.

B. Effect of Polymers

The results of the effect of nonionic polymers have been shown in Figure 1. It may be seen that adsorption decreases with the addition of PVA up to a certain level (100 ppm of PVA) and then attains a plateau. This is attributed to the strong affinity of this polymer for clay surfaces. In the presence of scleroglucan and

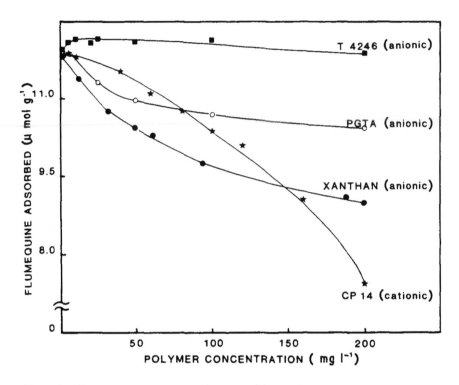

Figure 2. Effect of ionic polymers on adsorption of flumequine.

levan, adsorption reaches a minimum and then attains a plateau up to its original level (for levan) of adsorption and slightly higher (for scleroglucan) at an elevated polymer concentration. Scleroglucan is reported[5] to have strong affinity for clay surfaces. It also modifies the clay surface after adsorption. Thus, the initial decrease in the adsorption of flumequine in the presence of scleroglucan is due to the adsorption on clay by the latter substance. However, the subsequent increase results from the fact that higher concentration provides new adsorption sites for flumequine. It may be assumed that a similar effect (less intense) is true in the case of levan also. The other nonionic polymers (CMC and HP_8) showed little change in the adsorption of flumequine. This may be due to weak interactive forces between these polymers and clay surfaces.

The results of the effect of ionic polymers on adsorption of flumequine are shown in Figure 2. All the ionic polymers except T_{4246} showed decrease in adsorption. The decrease is only up to a level (100 ppm), and then it reaches a plateau at higher concentration in the case of polygalacturonic acid (PGTA). In the presence of CP_{14} and xanthan, the decrease in adsorption continues even at a higher concentration. It has been reported[6] that CP_{14} adsorbs strongly on the clay surfaces due to the electrostatic attractions and causes the formation of large flocs. The continued decrease in the adsorption of flumequine may be attributed to this flocculation effect. In the case of two anionic polymers (PGTA and xanthan), the decrease in adsorption is due to competition for the adsorption site. The continued decrease in adsorption even at a higher concentration results from its strong interactions with clay surfaces.

In the case of T_{4246}, the adsorption of flumequine slightly increases up to a level (10 ppm of T_{4246}) and then reaches a plateau. Because this polymer is reported[6] to stabilize the clay suspension, the increase in adsorption of flumequine may be due to this effect.

IV. CONCLUSIONS

The following conclusions are drawn from the present study:

1. Nonionic polymers, except PVA and scleroglucan, have little effect on the adsorption of flumequine. PVA caused decrease in adsorption at a low polymer concentration and then reached a plateau. Scleroglucan and levan showed a minimum in adsorption and reached a plateau up to the original level of adsorption (levan) or slightly higher (scleroglucan).
2. Anionic polymers, except T_{4246}, competed with ionized flumequine.
3. The decrease in adsorption continues even at a higher concentration of cationic polymer (CP_{14}) due to the flocculation effect.
4. Competition is stronger in the case of xanthan than in the case of PGTA, as the adsorption continues to decrease even at higher concentrations of xanthan.

ACKNOWLEDGMENTS

The authors wish to thank Professor H. Doner (University of California, U.S.) and Dr. C. Chenu (Station de Science du Sol, Versailles, France) for their valuable suggestions in this work and for providing certain polymer samples. We also thank Mr. Chevalier (3M/Riker, France) for providing labeled as well as unlabeled samples of flumequine.

REFERENCES

1. Kerhoas, L. and J.C. Thoisy-Dur. "Relation entre Adsorption et Persistence d'un Pesticide dans le sol. Etude du Comportement de la Flumequine," in *Methodological Aspects of the Study of Pesticides Behaviour in Soil*, P. Jamet, Ed. (Versailles, France: INRA, 1988), pp. 195–200.
2. Thoisy-Dur, J.C., M. Djafer, M. Terce, and M. Tabareau. "Etude de l'adsorption de la Fluméquine par un oxyde de fer: rôle du pH," in *Methodological Aspects of the Study of Pesticides Behaviour in Soil*, P. Jamet, Ed. (Versailles, France: INRA, 1988), pp. 37–47.

3. Khandal, R.K., J.C. Thoisy-Dur, and M. Terce. *Geoderma.* 50, 95–107 (1991).

4. Ben-Hur, M. and J. Letey. *Soil Sci. Soc. Am. J.* 53:233–238 (1989).

5. Chenu, C., C.H. Pons, and M. Robert. Proceedings of the International Clay Conference, Bloomington, IN, (1987), 375–381.

6. Aly, S.M. and J. Letey. Polymer and Water Quality Effects on Floculation of Montmorillonite. *Soil Sci. Soc. Am. J.* 52:1453–1458 (1988).

CHAPTER 10

Groundwater Contamination by Pesticides: Field Experiments in Shallow and Deeper Groundwater

U. Müller-Wegener, R. Schmidt, C. Ehrig, B. Ahlsdorf, and G. Milde

I. INTRODUCTION

The penetration of pesticides into the groundwater is influenced by a series of factors.[1,2,7,8] In this process, the physicochemical properties of the compounds used as well as their mobility and persistence play a decisive role.[3,4,6] In addition, the practice of application used in agriculture, i.e., frequency and date of application, the quantity used, and the technique of application are of importance.[9] As regional factors, the vulnerability of the groundwater has to be taken into account where geologic formation has an influence on the possible contamination of groundwater by pesticides as well as soil texture, clay content, and content of organic carbon in the soil.

The participation of the individual factors in the possible displacement of pesticides to the groundwater is shown in a series of investigations. The following experimental methods were applied:

1.   laboratory experiments
     - adsorption and desorption studies using selected active agents
     - percolation experiments in small columns under saturated and unsaturated flow conditions
     - degradation studies under aquifer-like conditions

0-87371-616-7/93/$0.00 + $.50

Table 1. Physicochemical Properties of the Active Agents

Active agent	Water solub.[a]	Vapor press.[a]	$t_{1/2}$[b]	K_{oc}[c]	Mobility in soils[a]
Aldicarb	3	2	10	35–46	Very mobile
Atrazine	2	0	10–20	90–110	Mobile-moderate mobile
Simazine	1	0	4–16	140	Mobile-moderate mobile
Chlortoluron	2	0	20–25	420	Moderate mobile

[a] Classified in five categories.
[b] Weeks in the upper soil layers.
[c] Sorption constant related to the content of organic carbon.

2. field studies
   - displacement experiments in lysimeters
   - field studies in shallow groundwater
   - field studies in deeper groundwater including wells of waterworks

II. ACTIVE AGENTS AND EXPERIMENTAL SITES

For the experiments, those active agents were selected from the approximately 280 agents approved for agricultural use which are of considerable importance due to their wide-spread use. Being applied also in special cultures in higher quantities, their analytical determination is easy. The active agents should show clear differences in their physicochemical behavior. Moreover, it should be possible to observe different periods of application. The selected active agents are listed in Table 1.

The sites for the field studies had to meet three requirements:

1. The pesticides selected should be applied intensively and on an area which is as large as possible.
2. They should include permeable soils with varying contents of organic carbon.
3. They should show a low depth of groundwater level.

The sandy soils of the heterogeneous loose sediments of northern Germany with extensive tree nursery cultures, e.g., in the neighborhood of Pinneberg, proved to be most suitable.

III. RESULTS AND DISCUSSION

For degradation of the pesticides investigated under approximate aquifer conditions (i.e., darkness, temperatures between 8 and 12°C, approximate anaerobic conditions, initial concentrations of approx. 2 μg/L), clear differences have been observed at a water/substrate ratio of approx. 1:1 (Figures 1 and 2).

For the evaluation of a possible degradation in the deeper soil, the considerably reduced degradation as compared to the upper soil should be taken into account.

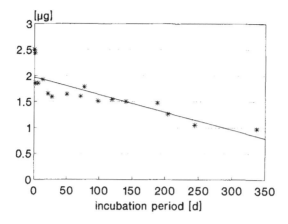

Figure 1. Decomposition of atrazine under aquifer-like conditions.

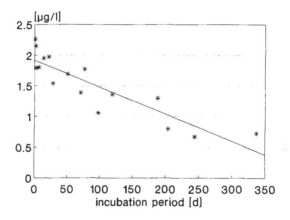

Figure 2. Decompositon of simazine under aquifer-like conditions.

Half-life periods of approx. 200 days for simazine and 300 days for atrazine are clearly above those given for the upper soil (approx. 30 days). Thus, the data obtained on degradation in the upper soil which is intensely inhabited cannot be applied even approximately to the degradation process in the area of the aquifer.

Similar experience has been gained in leaching experiments under saturated and unsaturated conditions. The experiments in the saturated area which can be performed rapidly and with little expenditure on equipment only reflect the water movement in the aquifer or the conditions of complete water saturation in the upper soil occurring very rarely. The unsaturated flow, however, is the normal situation; and in this case, the pesticides investigated show quite a different behavior. Thus, only half or even one third of the elution volume is necessary to achieve displacement of pesticides in the unsaturated flow, under otherwise comparable conditions.

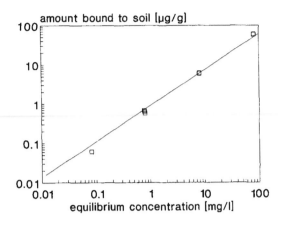

Figure 3. Adsorption of ^{14}C-labeled aldicarb to cambisol.

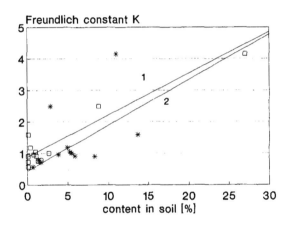

Figure 4. Adsorption of ^{14}C-labeled aldicarb to soils depending on the C_{org} (1,□) (r = 0.95) and clay content (2,*) (r = 0.507) of the soil samples.

Also in the case of adsorption studies, it is not possible to consider the behavior in a soil horizon as representative for the whole soil, as has been frequently done. An example for that is the adsorption of ^{14}C-labeled aldicarb to a cambisol (Braunerde) (Figure 3). There is a clear dependency of adsorption, represented by the constant K of the Freundlich equation, in the area of the upper soil on the content of organic carbon (Figure 4). In the deeper areas which are poor in humus, however, the content of clay minerals is the factor regulating adsorption.

Nevertheless, the correlation of clay contents with k values of the Freundlich equation is masked considerably by C_{org} contents so that a linear dependency for them is only slightly visible. The multiple regression shows, at an accuracy of $R^2 = 0.935$, that changes in the C_{org} content of the soil are, with a factor of

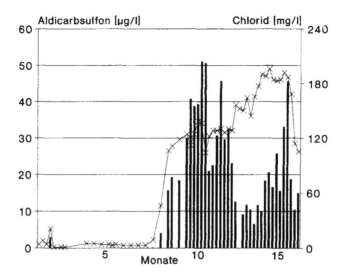

Figure 5. Lysimeter effluent 2 m below a layer of cambisol for chloride (x) and aldicarbsulfon after application of 5 kg/ha.

approx. 2.5, more clearly visible than those in the clay content ($k = 0.678 + 0.116 [C_{org}\%] + 0.0470 [clay\%]$).

Even if all parameters which have an influence on the results of these laboratory experiments are taken into account, it is possible to forecast the behavior of pesticides in the field on the basis of laboratory experiments in individual cases only. This is due to natural irregularities of the soil such as the uneven distribution of the pores, the occurrence of cracks caused by dryness, or the irregular distribution of rainfall.

When using lysimeters in the open air, natural variations are included to a greater extent in the evaluation of displacement of pesticides. Even if simple effluent curves are drawn up, the active agent may, e.g., be detected in the percolating water at a depth of 2 m as soon as one month after the application of 5 kg aldicarb/ha (Figure 5). This is due to the coarse pores of the soil which favor transportation. The actual displacement into the groundwater of aldicarb takes place with little delay to the waterfront (approx. 8 months after application).

In the lysimeter and field studies, instead of the active substance, aldicarb, a main soil metabolite, aldicarbsulfon, was detected. In the injector of the gas chromatograph (GC), aldicarbsulfonnitril was built by thermolysis. A separation of such nitril which develops in degradation processes in the soil itself, was not possible. Separate measurements of the nitril concentration present in the soil after degradation of aldicarb allowed an estimation of this part.

Results obtained from laboratory experiments may be confirmed by lysimeter experiments. Thus, soils with an increased content of organic carbon show a greater delay in the displacement to the groundwater as compared to the water-

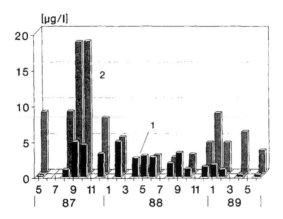

Figure 6. Course of simazine (1) concentrations (application of 1 kg/ha for 20 years) and aldicarbsulfon (2) concentrations (application of 4-kg aldicarb/ha for 4 years) in shallow groundwater (2 m) under sandy brown earth.

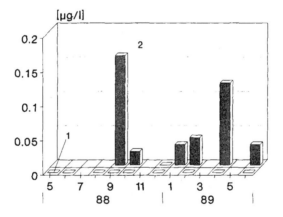

Figure 7. Simazine content in deeper groundwater (depth of the filter 10–22 m and 8–18 m) after application for 4 (1) and 20 (2) years at comparable sites.

front. A similar behavior has also been determined for clay contents if the C_{org} content of the soils investigated is low.

In the field studies, water samples were taken from tubes directly below the treated surfaces from different depths[5] and examined for the active agents applied. In these investigations, different aims were pursued.

In the first series, the course of concentrations of the active agents during the year were determined in shallow groundwater (approx. 2 m) of sandy soils (Figure 6). A long-term application of 1-kg simazine/ha led to clearly lower concentrations than a short-term application of aldicarb (4 kg/ha for 4 years). Moreover, the variations of concentrations during the year are considerably higher in the case of aldicarbosulfon, which is more mobile.

At two comparable sites, simazine could be clearly detected in deeper groundwater (up to approx. 20 m) after a period of application of 20 years (Figure 7).

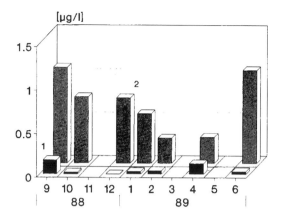

Figure 8. Simazine (1) and aldicarbsulfon (2) contents in deeper groundwater after a 4-year application of aldicarb and a 20-year application of simazine.

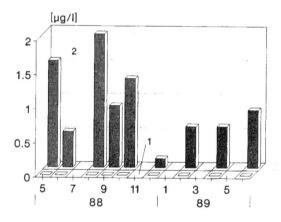

Figure 9. Concentrations of simazine and aldicarbsulfon after a 4- and 3-year application, respectively, in the catchment area of the well (filter depth 8–18 m).

An application period of 4 years did not yet lead to a contamination of the groundwater. Thus, the periods of regular use of the active agents are a decisive factor in the consideration of a possible contamination of the groundwater in question.

The fact that properties of the agents also play an important role may be seen from Figure 8. After the application of aldicarb for 4 years, concentrations of up to 1 µg/L could be detected even in the deeper groundwater. This very mobile agent penetrates into the lower soil considerably faster.

A direct comparison between aldicarb and simazine, applied to the surfaces for 3 and 4 years, respectively, is shown in Figure 9. While simazine could not be detected for 2 years, aldicarbsulfon could be found in concentratinos of up to 2 µg/L varying considerably in the course of the year.

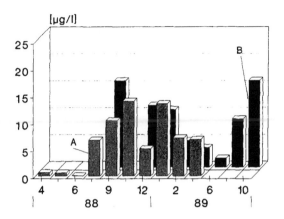

Figure 10. Concentrations of aldicarbsulfon (very mobile) below A: sand/loam/sand, depth of groundwater level 3 m, 4.7% organic substance; B: all sand, depth of groundwater level 3.6 m, 4.9% organic substance.

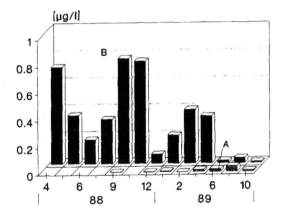

Figure 11. Concentrations of simazine (mobile-moderate-mobile) below A: sand/loam/sand, depth of groundwater level 3 m, 4.7% organic substance; B: all sand, depth of groundwater level 3.6 m, 4.9% organic substance.

Figures 10 and 11 show that mobility of the agents is of decisive importance in the evaluation of groundwater contamination. The effect of almost closed covering layers on penetration of the agents into the lower soil could be determined at two otherwise comparable sites. Almost equal concentrations of the very mobile aldicarb (Figure 10) could be detected in groundwater in spite of the protective loamy layer. The moerately mobile simazine, however, penetrates the loamy layer only in lowest concentrations.

Apart from the properties of the soil, the quantity of the agent applied and the period of application (primarily the properties of the active agent such as persistence and mobility) are decisive for the displacement of pesticides into the groundwater. Also when the aquifers are less vulnerable, very mobile agents may penetrate into the groundwater if applied in low quantities.

REFERENCES

1. Ahlsdorf, B., C. Ehrig, E. Zeeb, U. Müller-Wegener, and G. Milde. "Grundwasserbelastung durch Pflanzenschutzmittel, Felduntersuchungen in oberflächennahen und tieferen Grundwässern," *Mitt. Dtsch. Bodenkd. Ges.* 59:1033–1038 (1989).

2. Ahlsdorf, B., R. Stock, U. Müller-Wegener, and G. Milde. "Zum Verhalten ausgewählter Pflanzenschutzmittel in oberflächennahen Grundwässern heterogener Lockersedimente," in *Pflanzenschutzmittel und Grundwasser. Bestandsaufnahme, Verhinderungs- und Sanierungsstrategien,* G. Milde and U. Müller-Wegener, Eds. (Stuttgart: Gustav Fischer Verlag, 1989), pp. 375–395.

3. Bailey, G.W. and J.L. White. "Factors Influencing the Adsorption, Desorption, and Movement of Pesticides in Soil," *Res. Rev.* 32:29–29 (1970).

4. Capriel, P. and A. Haisch. "Persistenz von Atrazin und seiner Metaboliten im Boden nach einmaliger Herbizidanwendung," *Z. Pflanzenernähr. Bodenkd.* 146:474–480 (1983).

5. Friesel, P., R. Stock, B. Ahlsdorf, J. von Kunowski, B. Steiner, and G. Milde. *Untersuchung auf Grundwasserkontamination durch Pflanzenbehandlungsmittel. Materialien des Umweltbundesamtes,* (Berlin: Erich Schmidt Verlag, 1987).

6. Häfner, M. "Wichtige Aspekte zum Schutz des Grundwassers vor Pflanzenschutzmittel-Rückständen — dargestellt am Beispiel der Chlortriazine," in *Pflanzenschutzmittel und Grundwasser. Bestandsaufnahme, Verhinderungs- und Sanierungsstrategien* G. Milde and U. Müller-Wegener, Eds. (Stuttgart: Gustav Fischer Verlag, 1989), pp. 261–294.

7. Müller-Wegener, U., B. Ahlsdorf, N. Litz, and R. Stock. *Neue Erkenntnisse zum Auftreten von Pflanzenschutzmitteln in Grundwässern. Tätigkeitsbericht 1987 — Bundesgesundheitsamt* (Munich: MMV-Medizinverlag, 1988), pp. 91–92.

8. Müller-Wegener, U., C. Ehrig, B. Ahlsdorf, N. Litz, B. Katona, and G. Milde. "Zur Verlagerung von Pflanzenschutzmitteln in Böden," *Mitt. Dtsch. Bodenkd. Ges.* 59:433–438 (1989).

9. Müller-Wegener, U. and G. Milde. "Pflanzenschutzmittelanwendung und Grundwasserschutz — Eine Einführung zu den aktuellen Fragen," in *Pflanzenschutzmittel und Grundwasser. Bestandsaufnahme, Verhinderungs- und Sanierungsstrategien,* G. Milde and U. Müller-Wegener, Eds. (Stuttgart: Gustav Fischer Verlag, 1989), pp. 261–294.

CHAPTER **11**

Potential of Fluorescence Spectroscopy in the Study of Interactions of Pesticides with Natural Organic Matter

N. Senesi and T. M. Miano

ABSTRACT

The potential application of fluorescence quenching and fluorescence polarization techniques to the study of binding/adsorption of organic pollutants into natural organic matter is reviewed. Basic principles underlying the fluorescence phenomena as a whole and treatment methods of data obtained by fluorescence quenching and fluorescence polarization techniques are summarized. Experimental results and information available at present on the application of either fluorescence techniques to the study of interaction of polycyclic aromatic hydrocarbons (PAH) with humic substances (HS) are briefly discussed. Advantages provided by both fluorescence techniques in the quantitative and structural studies of organic pollutant-humic substance binding, possible extension and application to other organic pollutants (such as pesticides), and recommendations for further work conclude this review.

I. INTRODUCTION

A comprehensive model for predicting the transport and fate of organic xenobiotics, such as pesticides, in soils and aquatic systems should include both

0-87371-616-7/93/$0.00 + $.50

quantitative data based on kinetic and thermodynamic (equilibrium) measurements of biotic and abiotic processes that the pollutant molecules may undergo, and qualitative information on the speciation mechanisms and chemical forms in which the pollutant occurs in these systems. Within the abiotic compartment, adsorption processes are of concern in that they can substantially modify bioavailability, toxicity, mobility, and susceptibility to degradation of organic xenobiotics. With this respect, concentration and distribution in the environment of natural organic compounds, such as humic substances (HS), are ascertained to be of primary importance.[1]

Adsorption of organic pollutants is commonly studied by batch or dialysis equilibration with various initial sorbate:sorbent ratios. In the batch method, after a contact time sufficient to approach equilibrium, solids are separated from the mixture, generally by centrifugation or filtration on membranes or inert porous media; and equilibrium solution concentrations of organics are determined by suitable methods, such as gas and liquid chromatography, mass spectrometry, adsorption spectrometry, or radioactivity. The whole process is very time-consuming and involves multiple manipulations, imperfect solid/solution phase separation, possible compound degradation, and general analytical difficulties for trace organic analyses.

Conventional fluorescence spectroscopy and fluorescence polarization techniques have been applied successfully to the study of binding and/or adsorption of strongly fluorescent organic pollutants, such as polycyclic aromatic hydrocarbons (PAHs), onto natural organic matter, including humic substances (HS).[2-4] The potential extension of both fluorescence methods to the study of interactions of other environmentally important organic pollutants, such as pesticides, with fulvic acid (FA) and humic acid (HA) fractions of HS has been also emphasized.[5,6] To the knowledge of the authors, however, no information is available at present on studies in this direction.[7,8] Although HA and FA also exhibit an intrinsic weak fluorescence, they do not seem to interfere with fluorescence measurements of efficiently fluorescent organic pollutants.

The objective of this chapter is to briefly review basic principles underlying the fluorescence phenomenon as a whole and to discuss data treatment of fluorescence quenching and fluorescence polarization techniques, with particular reference to experimental results and information available at present on the application of either method to the study of PAH-HS interactive systems.

II. THE FLUORESCENCE PHENOMENON

The absorption of UV-Vis radiation by a molecule promotes electrons from the ground state to the excited state. Excited electrons return to the ground state by nonradiative (energy dissipated as heat) and radiative processes which include, among others, fluorescence, that is the emission of electromagnetic radiation of longer wavelength than the incident radiation.

Chromophoric structural entities containing conjugated double bonds and aromatic compounds (such as those present in numerous organic xenobiotics and HS) generally exhibit fluorescence, whereas relatively few aliphatic and alicyclic molecules are able to fluoresce. Numerous molecular and environmental factors, including the extent of the π electron system, nature of substituent groups, structural rigidity, steric hindrance, solvent, temperature, pH, presence of solutes and metal ions, etc., affect in various ways the fluorescence of a chromophore.[9,10]

Further, numerous phenomena including internal conversion, or collisional deactivation, intersystem crossing, photodecomposition processes, and inner-filter effects, compete or interfere to various extents with fluorescence emission of energy from excited electrons in a molecule.

The intensity of fluorescence, F_f, is given by:

$$F_f = \Phi_f F_o(1 - e^{-\epsilon bc}) \tag{1}$$

where Φ = quantum efficiency, i.e., the ratio of the total energy emitted as fluorescence per total energy absorbed

F_o = intensity of the incident radiation

ϵ = molar absorptivity at the excitation wavelength

b = path length of the cell

c = molar concentration

For very dilute, low-absorbing solutions, where ϵbc is sufficiently small, Equation 1 reduces to a linear relationship:

$$F_f = \Phi_f F_o \epsilon bc \tag{2}$$

When performing fluorescence measurements, calibration of spectrometer response with known fluorophores is necessary, because both the lamp intensity and detector response vary with wavelength and may produce spectral distortions and variations of instrumental sensitivity. In addition, corrections for background scattered light, which include Raman and coherent scattering recorded simultaneously with the fluorescence, are often necessary.

III. THE FLUORESCENCE QUENCHING METHOD

Various processes can reduce the intensity of emission of a fluorophore (quenching effect), a particular case being its association with another organic molecule or adsorption on organic or inorganic surfaces, such as colloidal clays.

The formation of a "simple" complex between a fluorescent organic pollutant, P, and a single type of site of a humic molecule, HS, may be represented by the following fundamental equations:

$$P + HS = P - HS \tag{3}$$

$$K_b = [P - HS]/[P][HS] \qquad (4)$$

The mass balance on P is described by:

$$[P]_{tot} = [P] + [P - HS] \qquad (5)$$

where $[P]_{tot}$ is the formal or total concentration of P in the system. Combining Equations 4 and 5 and rearranging, yields:

$$[P]_{tot}/[P] = 1 + K_b[HS] \qquad (6)$$

Assuming that the fluorescence intensity is proportional to the concentration of free P in solution, [P], then:

$$F_o/F = 1 + K_b[HS] \qquad (7)$$

where F_o/F = ratio of fluorescence intensities of the fluorescent pollutant, P, in the absence (F_0) and in the presence (F) of the quencher HS.

A Stern-Volmer plot can then be constructed for the dependence of F_0/F as a function of [HS] and further information on the mechanism(s) responsible for quenching, i.e., static or dynamic quenching, can be inferred.

The situation is different for complexes formed with homologous constituents, where HS and P-HS include a large number of complexing sites and complexes, respectively. In such cases, several complications arise and a rigorous interpretation of fluorescence data is much more difficult, with the estimation of only an average equilibrium quotient for the complexation being possible.

IV. APPLICATION OF FLUORESCENCE QUENCHING TO PAH-HS BINDING

McCarthy and Jimenez[3] analyzed the time course (rate) of binding of a PAH, benzo(a)pyrene (BaP), to a water-dissolved commercial HA by measuring the quenching of fluorescence emission at 405 nm of a solution of BaP excited at 380 nm, caused by addition of small volumes of HA solution and rapid mixing. Corrections of fluorescence data for the contribution of HA fluorescence and for the inner filter effect were minimal. The quenching of BaP fluorescence resulted in a decrease in the value of F relative to F_0 and was measured as an increase in the F_0/F ratio with the addition of HA (Figure 1).[3]

The quenching of BaP fluorescence by HA followed the classic Stern-Volmer relationship with the ratio F_0/F being linearly related to the concentration of the quencher (HA). At any concentration of HA, the quenching was complete within 5–10 min after the addition of HA (Figure 1). This suggested that the association of BaP with HA is very rapid; that is, the binding of BaP to HA exhibits only

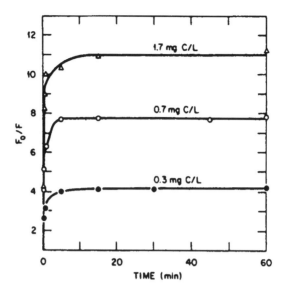

Figure 1. Time course of quenching of BaP fluorescence following the addition of HA at time zero. HA concentrations are indicated. F_0/F is the ratio of fluorescence in absence of HA to that in presence of HA. (From McCarthy, J.F. and B.D. Jimenez. *Environ. Sci. Technol.* 19:1072–1076 [1985]. With permission.)

a fast kinetic component and does not involve a slow component, as to that postulated in other systems. Fluorescence data combined with data obtained by using equilibrium dialysis showed, however, that dissociation of BaP from HA is equally reversible regardless of the association time. These results suggested that dissolved HS are an important factor to be considered in processes controlling the transport, fate, and biological effects of hydrophobic contaminants in the environment. They can, in fact, compete with other suspended or solid particles for the binding of contaminant and greatly reduce the contaminant availability for uptake and bioaccumulation by organisms. No attempt was made by Mc-Carthy and Jimenez[3] to directly quantify the extent of binding by fluorescence measurements.

A fluorescence quenching method has been developed successively by Gauthier et al.[4] based on the observation that PAHs fluoresce in aqueous solution in the free state but not when associated with HS. The method allowed the determination of equilibrium constants for the association of perylene, phenanthrene, and anthracene with water-dissolved HA and FA originated from two different podzol soils and a riverine water. Since the fraction of PAH associated with HS could be determined directly from the fractional decrease in fluorescence intensity (F_0/F) as a function of added HS, association constants (K_b) could be derived with Stern-Volmer plots on the basis of Equations 3–7.

The resulting Stern-Volmer plot of F_0/F values (fluorescence intensities were corrected for dilution effects and the inner filter effect) vs fulvic acid is shown

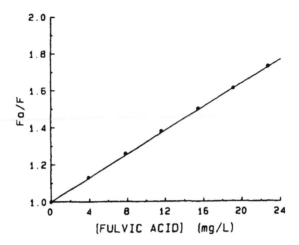

Figure 2. Stern-Volmer plot for F_o/F vs fulvic acid. The slope is 0.032 L/mg, the intercept is 1.00, and the value for R^2 is 0.999. (From Gauthier, T.D., E.C. Shane, W.F. Guerin, W.R. Seitz, and C.L. Grant. *Environ. Sci. Technol.* 20:1162–1166 [1986]. With permission.)

in Figure 2 for phenanthrene.[4] In all cases, the interaction of HA or FA with the three PAHs studied resulted in linear plots. Further analysis of the fluorescence data, in order to determine which type of mechanism was responsible for the quenching, excluded a dynamic or collisional quenching (i.e., nonradiative deactivation of the fluorophore by interaction with the quencher), and concluded in favor of a static quenching process, that is, the formation of a nonfluorescent complex between the fluorophore PAH and the HS quencher.[4]

To further confirm the validity of the fluorescence quenching technique,[4] determined K_{OC} values for the interaction of anthracene with the various HAs and FAs studied under the same pH and ionic strength conditions by using both fluorescence quenching and a reverse-phase separation technique. Although direct comparison of the two methods was intrinsically difficult, anthracene-humic association constants determined by the two methods correlated reasonably well. The fluorescence quenching technique is, however, only applicable to measure K_{OC} values for compounds with high fluorescence efficiencies, such as PAHs.

V. THE FLUORESCENCE POLARIZATION TECHNIQUE

Although the fluorescence polarization experiment has been widely used in biomedical sciences for the study of proteins[11,12] and in synthetic polymer sciences, it has been applied to the study of pollutant binding to environmental materials only in two cases.[2,13]

The fluorescence polarization technique is based on the measurement of the polarization of fluorescence emitted by a molecule that has been excited with plane polarized light. The polarization, p, is calculated by measuring fluorescence intensities in the planes parallel and perpendicular to the plane of polarization of the excitation radiation, denoted $I_\parallel$ and $I_\perp$, respectively:

$$p = \frac{I_\parallel - I_\perp}{I_\parallel + I_\perp} \qquad (8)$$

The main process which depolarizes fluorescence is molecular rotation during the lifetime of the excited state, which causes a decrease in the measured value of p. Thus, if the fluorophore undergoes significant rotation during the lifetime of the excited state, fluorescence will be depolarized, i.e., the measured value of p will be less than expected. Any process which causes a change in the rate of rotation of a fluorophore (e.g., adsorption on suspended particulate matter or binding to dissolved materials, both involving a variation in size of the fluorophore) will produce a change in polarization and can be followed by polarization measurements. The polarization for the bound, or adsorbed, fluorophore is higher than that for the free fluorophore because its effective size in the bound/adsorbed form increases and it rotates more slowly.

By measuring values of polarization for the free/dissolved fluorophore (p_f) and for the bound/adsorbed fluorophore (p_b) it is possible to relate the observed value of p in the mixture of free and bound fluorophores to the fraction of fluorophore adsorbed (F_b/F_f), that is, the relative amount (or ratio of concentrations) of bound and free fluorophore.[6] The basic equation is:

$$F_b/F_f = (Q_f/Q_b)[(p - p_f)/(p_b - p)] \qquad (9)$$

where Q_f/Q_b is the ratio of fluorescence efficiencies for free and bound fluorophore. Q_f and p_f are readily determined from a solution of dissolved fluorophore in the absence of sorbent material. Q_b and p_b are measured in the presence of a large excess of sorbent material, so that essentially all the fluorophore is adsorbed. Once these values have been determined, data for a complete adsorption isotherm can be acquired by titrating the binding or sorbent material into a fluorophore solution at constant total concentration and by measuring polarization, p, as a function of added material. Alternatively, the sorbent material concentration, or amount, may be held constant and the amount of fluorophore added varied. Thus, sufficient information is available to calculate an equilibrium constant for binding, if stoichiometry is known or can be assumed.

Relative fluorescence intensities for bound and free fluorophore can be used for the determination of Q_f/Q_b ratio, provided that fluorescence intensity is not affected by inner-filter effects associated with absorption or interference by scattering due to turbidity. Otherwise, a correction factor can be established and applied. Other problems may be encountered, for example, with the experimental possibility of distinguishing p from p_b or p_f, if the binding is very strong (>99% of fluorophore is bound) or very weak, respectively. These problems can be overcome by achieving suitable experimental conditions, unless binding is unusually strong or unusually weak.

The absolute polarization intensities, $I_\parallel$ and $I_\perp$, provide additional information as it concerns eventual effects on the adsorption measurement such as fluorophore degradation processes. Changes in absolute fluorescence intensity with time indicate, in fact, that fluorophore concentration is changing.

VI. APPLICATION OF FLUORESCENCE POLARIZATION TO PAH-FA BINDING

Although the potential of fluorescence polarization as a valuable means for studying sorption of organics on a variety of sorbent materials including HS has been emphasized since 1982 by Grant and Seitz,[6] only one study is available on the application of this technique to the interaction of a podzol soil FA to perylene, an efficient fluorophore representative of the PAH micropollutant class.[2]

Roemelt and Seitz[2] found that fluorescence of perylene in 75% glycerol solutions at pH values 2, 7, or 11 becomes more polarized as a function of added FA (Figure 3). These results indicated that perylene binds to FA to form a larger species that rotates more slowly in solution than does perylene alone. Perylene polarization reached a maximum value and then remained constant at FA levels above $3 \times 10^{-6} M$, thus indicating that the binding is complete at this concentration. The initial polarization value (no FA added) is p_f, and the final value is p_b. The relatively small overall change in polarization suggested that the reaction involves one perylene molecule per one FA molecule, i.e., a 1:1 stoichiometry could be reasonably assumed. If a larger species was formed, a larger change in polarization would be expected. Since the fluorescence intensity from perylene is not affected by added FA, the Q_b/Q_f term can be considered equal to 1.

Based on these assumptions, Roemelt and Seitz[2] calculated a binding equilibrium constant for the perylene-FA reaction:

$$\text{pery} + \text{FA} = \text{pery} - \text{FA} \tag{10}$$

directly from polarization data by derivation of Equation 9:

$$K_b = [\text{pery} - \text{FA}]/[\text{pery}][\text{FA}] = (p - p_f)/\{(p_b - p) \tag{11}$$

$$\{C_{FA} - [(p - p_f)/(p_b - p)] C_{pery}\}\}$$

where C_{FA} and C_{pery} are the total concentrations of FA and perylene, respectively. The binding constants calculated from the point where $p - p_f = p_b - p$ (i.e., the point where the greatest precision is expected in the calculation) were 1.2×10^6, 1.8×10^6, and 1.5×10^6 at pH 2, 7, and 11, respectively.

Important structural information on the perylene-FA interaction could be inferred from these results. The quite large values calculated for binding constants

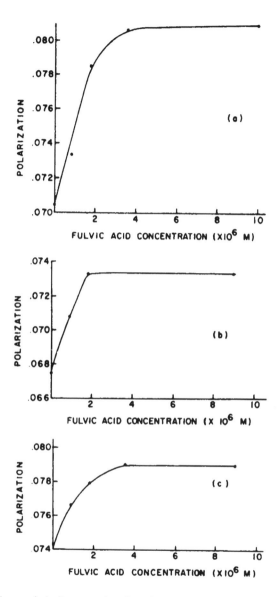

Figure 3. Perylene polarization as a function of added fulvic acid in 75% glycerol (4.9 × 10⁻⁷ *M* perylene): (a) pH 2; (b) pH 7; (c) pH 11. (From Roemelt, P.M. and W.R. Seitz. *Environ. Sci. Technol.* 16:613–616 [1982]. With permission.)

in 75% glycerol solution, which are expected to be considerably larger in water, confirmed the existence of large hydrophobic surface regions in FA available for association to highly hydrophobic organic pollutants (such as perylene), and suggested that this process is likely to be significant in natural systems. The similarity of binding constant values measured at different pHs would indicate that the extent of ionization of carboxylic and phenolic OH groups does not affect the ability of hydrophobic regions of FA to associate with hydropobic organics. Further, the fact that the greatest change in perylene polarization on binding to FA was observed at pH 2 and the smallest change occurred at pH 11 is consistent with a different conformation of the FA molecule in solutions at different pHs.

VII. ADVANTAGES AND EXTENSION OF FLUORESCENCE TECHNIQUES

Both the fluorescence quenching method and fluorescence polarization technique, where applicable, appear particularly attractive for offering significant advantages in quantitative and molecular studies of the binding/sorption of organic pollutants onto natural organic matter, notably humic substances. The major advantage of both techniques is that no separation step is required, thus eliminating possible errors due to incomplete separation of free/dissolved from bound/sorbed pollutant, with consequent potential improvements in reproducibility. Both techniques qualify as relatively rapid and convenient and offer good precision and high inherent sensitivity, thus enabling measurements at low pollutant concentration at environmental levels.

The fluorescence polarization technique is preferentially applied where the pollutant molecule is an efficient fluorophore with respect to the HS molecule. The fluorescence quenching method represents an important alternative to the polarization technique in systems where the fluorescence of pollutant is affected by the presence of HS.

Measurements based on changes of intrinsic fluorescence polarization of the humic molecule could extend this technique to nonfluorescent organic pollutants. Since the humic molecule is typically a nonefficient fluorophore and is large relative to most pollutants, the polarization changes to be measured are likely to be small and difficult to measure with precision. This will, therefore, require a high quality instrumentation designed for precise, sensitive polarization measurements.

In principle, the fluorescence polarization method is not restricted to fluorescent compounds, in that it can be extended to nonfluorescent compounds by competitive binding experiments, where known amounts of fluorophore probe and nonfluorescent compound-of-interest compete for a limited number of binding sites. This approach has been successfully applied to studies of protein binding to small molecules.[11] The binding constant for adsorption of the nonfluorescent compound thus can be calculated from knowledge of the level to which the fluorophore is displaced by the nonfluorescent compound and of the binding constant of the fluorophore, provided that both species compete for the same binding sites. Irreversible binding and slow kinetics, however, could complicate this kind of experiments.

REFERENCES

1. Senesi, N. and Y. Chen. "Interactions of Toxic Organic Chemicals with Humic Substances," in *Toxic Organic Chemicals in Porous Media,* Z. Gerstl et al., Eds. (Berlin: Springer-Verlag, 1989), pp. 37–90.
2. Roemelt, P.M. and W.R. Seitz. "Fluorescence Polarization Studies of Perylene-Fulvic Acid Binding," *Environ. Sci. Technol.* 16:613–616 (1982).
3. McCarthy, J.F. and B.D. Jimenez. "Interactions Between Polycyclic Aromatic Hydrocarbons and Dissolved Humic Material: Binding and Dissociation," *Environ. Sci. Technol.* 19:1072–1076 (1985).
4. Gauthier, T.D., E.C. Shane, W.F. Guerin, W.R. Seitz, and C.L. Grant. "Fluorescence Quenching Method for Determining Equilibrium Constants for Polycyclic Aromatic Hydrocarbons Binding to Dissolved Humic Materials," *Environ. Sci. Technol.* 20:1162–1166 (1986).
5. Seitz, W.R. "Fluorescence Methods for Studying Speciation of Pollutants in Water," *Trends Anal. Chem.* 1:79–83 (1981).
6. Grant, C.L. and W.R. Seitz. "The Potential of Fluorescence Polarization for Measuring Sorption Isotherms of Organics," *Soil Sci.* 133:289–294 (1982).
7. Senesi, N. "Molecular and Quantitative Aspects of the Chemistry of Fulvic Acid and Its Interactions with Metal Ions and Organic Chemicals. Part II. The Fluorescence Spectroscopy Approach" *Anal. Chim. Acta* 232:77–106 (1990).
8. Senesi, N. "Fluorescence Spectroscopy Applied to the Study of Humic Substances from Soil and Soil Related Systems: A Review," 199th American Chemical Society Meeting, Boston, Div. Environ. Chem., Vol. 30, n.1, (1990), pp. 79–82.
9. Becker, R.S. *Theory and Interpretation of Fluorescence and Phosphorescence* (New York: Wiley-Interscience, 1969).
10. Guibault, G.G. *Practical Fluorescence. Theory, Methods and Techniques* (New York: Marcel Dekker, 1973).
11. Dandliker, W.B., R.J. Kelly, J. Dandliker, J. Farquhar, and J. Levin. "Fluorescence Polarization Immunoassay. Theory and Experimental Method," *Immunochemistry* 10:219–227 (1973).
12. Dandliker, W.B., J. Dandliker, S.A. Levison, R.J. Kelly, A.N. Hicks, and J.U. White. "Fluorescence Methods for Measuring Reaction Equilibria and Kinetics," in *Methods of Enzymology,* Vol. 48, C.H.W. Hirs and S.N. Timasheff, Eds. (New York: Academic Press, 1978), pp. 380–415.
13. Von Wandruszka, R.M.A. and S. Brantley. "A Fluorescence Polarization Study of Polyaromatic Hydrocarbons Adsorbed on Colloidal Kaolin," *Anal. Lett.* 12:1111–1122 (1979).

CHAPTER 12

Retention of Atrazine by Humic Substances of Different Natures

A. Piccolo and G. Celano

I. INTRODUCTION

It has been well documented that organic matter plays a major role in the adsorption of herbicides in soils and that organic matter content is usually the soil factor most directly related to herbicidal dynamics.[1] Despite this widely accepted principle, only a few works have attempted to correlate the chemical and physical-chemical characteristics of humic substances, the bulk of soil organic matter, with herbicidal behavior in soil.[2] Most of the literature is concerned with absorption of herbicides on whole soils failing to appreciate the variability introduced by the different nature of humic substances.

The objective of the work reported here was to study the absorption and desorption of atrazine on three different purified humic substances before and after acidic hydrolysis.

II. MATERIALS AND METHODS

A. Humic Substances

Humic substances were extracted from: (1) a volcanic soil (typic dystrandepts); (2) a North Dakota leonardite (Mammoth, Chem. Co.); and (3) an oxidized coal

0-87371-616-7/93/$0.00 + $.50

123

(Eniricerche, SpA). Extraction was with an alkaline solution under N_2 by the procedure outlined by Stevenson.[1] The resulting humic acids were purified by a HCl-HF treatment,[3] dialyzed against water, and freeze-dried. Part of these samples underwent acidic hydrolysis with 6 N HCl overnight, were again dialyzed, and then freeze-dried.

Total acidity of humic samples were determined by the calcium acetate method; phenolic groups were obtained by subtracting carboxyls from total acidity.[4] Molecular weight distribution was achieved by eluting a 20-mg sample at constant flow with 1 M 2-amino-2-hydroxymethyl-1,3-propanediol (TRIS) hydrochloride buffer at pH = 9 on a Sephadex G150 column (70 × 1.6 cm). Eluate absorbance (470 nm) was automatically recorded with a continuous flow spectrophotometer ISCO UA-5. Solution spectra of carbon 13 NMR of the purified humic acids were obtained in a quantitative mode using a Varian XL 300 spectrometer with inverse-gated decoupling, 450 pulse, acquisition time of 0.1 sec, and relaxation delay (decoupler off) of 1.9 sec.[5] Infrared (IR) spectra were recorded on a Perkin-Elmer 882 IR spectrometer by the KBr pellet technique.

B. Atrazine Interaction with Humic Substances

1. Adsorption and Desorption Kinetics

A 20 mL solution of 25 ppm of pure atrazine (99%) in 0.01 M $CaCl_2$ was shaken with 100 mg of humic sample at 25°C for four different times of contact (1, 2, 4, and 16 hr) and centrifuged; then the amount of atrazine left in solution was determined. Most of the supernatant was removed, the volume was measured, and 20 mL of 0.01 $CaCl_2$ was added to the humic residue. The suspension was shaken (1, 2, 4, and 16 hr) and centrifuged; then the amount of atrazine in solution was determined (first desorption). This procedure was repeated once (second desorption).

2. Mass Extraction and Adsorption Isotherms

A 100-mg sample was shaken for 16 hr with 20 mL of a 25-ppm solution of atrazine in 0.01 M $CaCl_2$ and centrifuged; then the supernatant was measured and discarded. To the residue 20 mL of a mixture of $CH_3CN:H_2O$ (9:1) was added and shaken for 16 hr. Next the suspension was centrifuged and the atrazine in the supernatant determined (first desorption). The procedure was repeated again (second desorption). Adsorption isotherms were constructed by letting 100 mg of humic sample be in contact with 20 mL of atrazine solution in $CaCl_2$ of increasing concentrations (1, 5, 10, and 25 ppm) at different contact times (1, 2, 4, and 16 hr).

The atrazine left in solution and that adsorbed per unit weight of humic sample were fitted into the logarithmic form of the Freundlich empirical equation, and the adsorption constant (K_f) and the slope of the isotherms (1/n) were obtained.

Table 1. Percent Adsorption of Atrazine per Unit Weight of Humus

Contact h	1	2	4	16
HA1	24.98	26.78	28.33	31.76
HA1-H	61.22	69.68	75.40	69.17
HA2	27.31	24.25	35.00	45.26
HA2-H	26.72	28.21	30.74	45.64
HA3	48.68	62.20	68.52	84.06
HA3-H	5.23	11.93	17.20	26.97

Table 2. Atrazine Percent Desorption at Four Contact Times and Mass Extraction (ME) at 16 hr per Unit Weight of Humus

Contact h	1	2	4	16	ME
HA1	71.09	71.90	73.81	65.23	80.16
HA1-H	44.72	25.83	21.55	34.07	77.17
HA2	74.97	91.75	81.81	64.25	87.27
HA2-H	44.72	27.19	28.45	38.38	85.19
HA3	21.04	22.60	25.14	14.23	69.91
HA3-H	42.22	29.32	38.96	37.55	95.62

C. Atrazine Determination

Atrazine was determined by high-performance liquid chromatography (HPLC) using a C18 reverse-phase Spheri5 Brownlee 220- × 4.6-mm column, propachlor as the internal standard, an isocratic elution of 60% $CH_3 CN$ in H_2O at 2 mL/min, and an UV (220-mm) detector.

III. RESULTS AND DISCUSSION

Kinetics of atrazine adsorption on the humic extracts (Table 1) show that percent adsorption on soil (HA1) and leonardite (HA2) material did not differ within 2 hr of contact, but it increases more rapidly for HA2 than for HA1 going toward 16 hr of contact. Much higher adsorption is shown by the humic material from oxidized coal which retains almost 49% of atrazine after only 1 hr of contact and reaches 84% adsorption after 16 hr of contact time.

Acidic hydrolysis brings about a rather drastic change of adsorption on HA1 and HA3 whereas the behavior of HA2 does not vary significantly. Hydrolyzed HA1 becomes the strongest adsorber with almost three times as much more atrazine than the original sample and an adsorption maximum at 4 hr of contact time. Conversely, the hydrolyzed extract from oxidized coal (HA3-H) lost most of its retaining capacity, reaching only about 5% adsorption after 1 hr of contact and a maximum of no more than 27% after 16 hr of contact time.

Kinetics of atrazine desorption seemed to follow the same pattern as shown by the adsorption kinetics (Table 2). Percent desorption was very high for HA1 and HA2 showing not only that adsorption was limited on these materials but

Table 3. Values for K_f and 1/n from Adsorption Isotherms Calculated from Logarithmic Form of the Freundlich Equation at Different Contact Times

Contact h	1		2		4		16	
	K_f	1/n	K_f	1/n	K_f	1/n	K_f	1/n
HA1	2.16	0.81	2.16	0.83	2.18	0.89	2.39	0.89
HA1-H	2.87	0.70	2.96	0.66	2.93	0.70	3.13	0.66
HA2	2.15	0.94	2.21	0.90	2.29	0.92	2.63	0.85
HA2-H	2.69	0.71	2.66	0.59	2.54	0.73	2.59	0.81
HA3	2.51	0.81	2.69	0.73	2.79	0.76	3.02	0.69
HA3-H	2.01	0.72	2.81	0.63	2.89	0.63	3.02	0.57

also that the adsorbed atrazine is not strongly bound. Desorption decreased significantly for the hydrolyzed products of HA1 and HA2. Surprisingly, though the former had shown a percent adsorption similar to both HA2 and HA2-H, its desorption was significantly less for the hydrolyzed product suggesting that adsorption sites of higher energy are available after hydrolysis. Desorption for HA3 reflects the adsorption behavior since less atrazine is desorbed from the high adsorbing original extract whereas desorption is higher (though comparable to HA1-H and HA2-2) from the hydrolyzed product. Mass extraction also confirms the differences in atrazine adsorption capacity of HAs. Extractability from HA1 and HA2 does not change significantly passing from the extract to its hydrolyzed product; whereas for HA3, mass extraction is about 25% more efficient than for HA3-H, substantiating the strength of atrazine adsorption on this material.

Adsorption isotherms constructed through the Freundlich equation allow calculation of the adsorption constant, K_f, which is a useful index of the degree of adsorption; and 1/n, which is related to the number of adsorption sites of different surface energy (Table 3). The K_f increases with contact time with all original humic extract, those of HA3 being the highest values. Conversely, the hydrolyzed products showed a decrease of K_f for HA3-H at 1 hr of contact time and an enhancement for HA1-H and HA2-H for all contact times. Values for 1/n, which are comparable for HA1 and HA2, decrease somewhat for HA3. A sensible reduction of 1/n is observed in all hydrolyzed products, though to a larger extent for HA1 and HA2. It also is noted that the larger the increase in K_f is, the greater the 1/n decrease is. A regression line calculated for K_f and 1/n values produced a correlation coefficient (r) of 0.878. This result generally suggests that a high degree of adsorption (K_f) corresponds to a flattening of the slope of Freundlich adsorption isotherms which must indicate an increase of the bonding surface energy.

Carbon 13 NMR spectra were obtained for the three humic extracts (Table 4) but not for the hydrolyzed materials due to instrument unavailability. The aliphatic carbon region (0–110 ppm) was the largest in HA1 followed by HA2 and HA3, in that order, whereas the extract from oxidized coal revealed the largest content of aromatic carbon (110–160 ppm). Carboxylic carbon was of the same order of magnitude in all samples though slightly higher in HA2 from leonardite.

Table 4. Percent Distribution of Carbon 13 Groups in Humic Materials as Measured by NMR

	Aliphatic (0–110 ppm)	Aromatic (110–160 ppm)	Carboxylic (160–190 ppm)
HA1	53.6	37.0	8.4
HA2	29.8	55.7	14.5
HA3	18.0	72.0	10.0

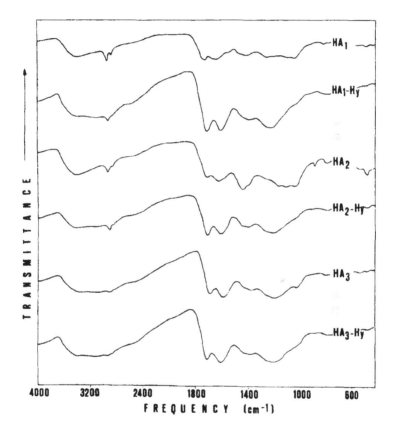

Figure 1. Infrared spectra of original and hydrolyzed humic materials.

Infrared spectra of original and hydrolyzed materials (Figure 1) were not able to reveal changes in structure except for an increased band at 1720/cm (COOH groups) in all spectra, probably due to the release of carboxyl groups involved in salt bridges prior to the acidic hydrolysis. A loss of aliphatic material (about 2900/cm) after hydrolysis was noticed for the humic acids extracted from the volcanic soil.

Chemical analysis showed (Table 5) that for all samples acidic hydrolysis produced a higher content of total and phenolic acidity whereas carboxylic acidity appeared reduced. Phenols may have originated from the hydrolysis of ester

Table 5. Content (meq/g) of
          Acidic Functional Groups of
          Humic Materials

	Total	Carboxylic	Phenolic
HA1	10.73	4.68	6.05
HA1-H1	3.35	4.11	9.24
HA2	9.30	4.39	4.92
HA2-H	10.09	3.72	6.37
HA3	9.02	3.58	5.44
HA3-H	11.63	3.39	8.24

bonds whereas carboxyl groups may have been lost through decarboxylation. Total acidity is associated with one of the adsorption mechanisms between atrazine and humic substances since the herbicide, being a weak base, can be protonated by the acidic functional groups of humic materials and thus adsorbed.[6] Variations found for the acidic groups (Table 5) before and after hydrolysis are generally similar for all three humic materials and do not explain the differences observed in adsorption and desorption of atrazine among the humic samples.

Conversely, variations between original and hydrolyzed materials are observed in molecular weight distribution (Figures 2, 3, and 4). Humic acid from soil showed, after hydrolysis, a large increase in absorbance for both eluted (high molecular weight) peaks suggesting that an enhanced concentration of light-absorbing groups (chromophores) in the macromolecular structure. Furthermore, a distinct shift of both peaks toward lower eluting volumes is observed, thereby indicating that acidic hydrolysis had condensed this HA into higher molecular weight. Less variation is observed for the humic acid extracted from leonardite (HA2) for which absorbance increased only for the high molecular weight peak, whereas the diffused peak is just slightly shifted toward higher molecular weights. The chromatograms of HA3 show that the material is mainly composed of a high molecular weight fraction eluting at the void volume, and hydrolysis decreases the condensation shifting the eluting peak toward low molecular weight. These results may partly explain the different behavior of humic materials in adsorbing and desorbing atrazine.

Concentration of chromophores and molecular dimension seemed to control interactions with atrazine. HA1 and HA2, which both showed a small peak for high molecular weight fraction with an absorbance never exceeding 0.36, gave comparable atrazine adsorption (Table 1) up to 2 hr of contact time. HA2, which revealed a diffused peak with absorbance about twice as much (0.7) as HA1 (0.4), was able to adsorb more than HA1 at higher contact times. The highest atrazine adsorber, HA3, was composed mainly by a high molecular weight fraction whose peak absorbance reached 1.18. The atrazine adsorbing capacity increased steadily with contact times up to 84%. Acidic hydrolysis drastically changed the chromatogram of HA1 by increasing the first peak absorbance (0.92) to almost HA3 levels as well as the percent of high molecular weight in the material. HA2 also showed an increase of the first peak absorbance, but to a much lower extent (0.46); and the diffused peak was slightly shifted to higher

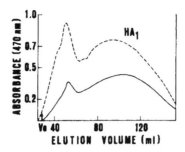

Figure 2. Molecular weight distribution for original (black line) and hydrolyzed (dotted line) humic acid from volcanic soil.

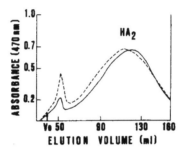

Figure 3. Molecular weight distribution for original (black line) and hydrolyzed (dotted line) humic acid from leonardite.

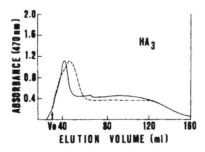

Figure 4. Molecular weight distribution of original (black like) and hydrolyzed (dotted line) humic acid from oxidized coal.

molecular weight. The change for HA3 was limited to a distinct shift of the first peak toward lower molecular weight. The strong absorbance and molecular weight increase of HA1 corresponded to the observed large enhancement of atrazine adsorption (Table 1). For HA2, the relative small changes introduced by hydrolysis in molecular weight distribution did not influence adsorption significantly. Conversely, in HA3 the reduction in molecular dimension, probably caused by the depolymerization following the hydrolysis of ester bonds, is related to the drastic decrease in atrazine adsorption.

Desorption kinetics seemed to respond to similar logic. The low molecular dimension and chromophore-poor HA1 and HA2 gave the largest desorption (Table 2); whereas H3, the highest adsorber, showed the lowest desorption thereby indicating the presence of high-energy adsorbing sites. Furthermore, desorption percent decreased substantially for the hydrolyzed HA1, reflecting the acquired adsorbing capacity of this material; and also for HA2 which, contrary to HA1, did not show a significant increase of atrazine adsorption after hydrolysis. This suggests that the change in HA2 molecular dimension provoked by acid hydrolysis was able not only to decrease its previously large adsorption capacity but also to reduce the bonding energy of adsorption sites as can be inferred by the enhancement of percent desorption.

IV. CONCLUSIONS

Carbon 13 NMR spectra of humic extracts have shown that the degree of aromaticity increases in the order: HA1 < HA2 < HA3. Previous results[5] have determined by means of carbon 13 NMR that acid treated humic substances lose preferentially aliphatic carbon components while they generally gain in degree of aromaticity. The increase in absorbance of gel permeation peak of HA1, originally rich in aliphatic components, observed with hydrolysis may well be interpreted as a loss of aliphatic carbon. The consequent concentration of aromaticity was also accompanied by an increase in molecular dimension. A similar behavior is shown by HA2, though to a lesser extent than one might have expected considering the lower content of aliphatic components. Conversely, HA3, which was mainly aromatic in nature, did not show variation in the first peak absorbance with hydrolysis; instead it showed a decrease of molecular dimension.

These results support a charge-transfer mechanism for atrazine adsorption on humic substances which is more effective the higher the molecular complexity of humic materials. A charge-transfer interaction may arise between electron-poor groups in humic acids (such as quinones) and in electron-rich groups in the atrazine molecule (such as the various nitrogen atoms).[6] Ion exchange and hydrogen-bonding mechanisms, though they may play a role, do not appear to command the adsorbing capacity of humic substances. Despite the increase for all samples in total carboxyl, and phenolic acidities observed after hydrolysis that should have driven towards a general higher atrazine adsorption, had protonation been the predominant mechanism, only HA1 showed an increase in adsorption. Our results are more appropriately explained by a charge-transfer mechanism with efficiency that is strengthened the higher the concentration of donor-acceptor groups and the larger the molecular complexity of humic materials. Senesi and Testini[7] based on electron spin resonance (ESR) results had already expressed doubts about ionic exchange in the leading mechanism of interaction between humic substances and s-triazines. The electron-donor processes that generate charge transfer bonds are then favored by high molecular dimensions probably for the formation of conjugated systems in which electrons could be shared, thereby enlarging the bonding surface energy.

REFERENCES

1. Stevenson, F.J. *Humus Chemistry: Genesis, Composition, Reactions* (New York: John Wiley & Sons, Inc., 1982).
2. Khan, S.U. *Pesticides in the Soil Environment* (Amsterdam: Elsevier, 1980).
3. Schnitzer, M. "Humic Substances: Chemistry and Reactions," in *Soil Organic Matter,* M. Schnitzer and S.U. Khan, Eds. (Amsterdam: Elsevier, 1978).
4. Schnitzer, M. "Organic Matter Characterization," in *Methods of Soil Analysis,* Part II, 2nd ed., Page et al., Eds. (Madison, WI: American Society of Agronomy, 1982), pp. 581–594.
5. Piccolo, A., L. Campanella, and B.M. Petronio. "Carbon-13 NMR spectra of soil humic substances extracted by different mechanisms," *Soil Sci. Soc. Am. J.* 54:750–756 (1990).
6. Stevenson, F.J. "Organic matter Reactions Involving Herbicides Analysis in Soil," *J. Environ. Qual.* 1:333–343 (1972).
7. Senesi, N. and C. Testini. "Physico-Chemical Investigations of Interaction Mechanism Between s-Triazine Herbicides and Soil Humic Acids," *Geoderma* 28:129–146 (1982).

CHAPTER 13

The Binding of Pesticide Residues to Natural Organic Matter, Their Movement, and Their Bioavailability

Francis Andreux, Irene Scheunert, Philippe Adrian, and Michel Schiavon

I. INTRODUCTION

Plant decay products and soil humus are the main sources of organic compounds dissolved in continental surface waters. For several decades, the study of transfer of solutions between soils and surface or groundwaters was limited to natural compounds, in close relationship with that of the physicochemical properties of the soils. Up to now, much literature about the numerous and interlinked soil processes of fluid dynamics in porous spaces, adhesion of mineral particles and microbial cells, rock weathering, metal complexation, and sorption of organic molecules[1-4] has been produced.

The increasing use of soils for the disposal of chemical refuses, as well as the generalized use of manufactured chemicals for several purposes in agricultural soils, has stimulated interest in the study of the fate of extraneous materials in natural ecosystems. For instance, of the 63×10^6 and 250×10^6 metric tons of organics that were released on the whole earth in 1970 and 1985, respectively, more than 1% were active pesticide molecules.[5] Pesticides represent a kind of *intentional pollution*, which was estimated in France to be 4.5 kg/ha of the agricultural surface area. It is now well established that a variable proportion of these pesticides and of their transformation products are not always quickly degraded, but interact with soil constituents, which can modify their mobility in profiles and landscapes. In relation with natural factors, the chemical structure of these pesticides determines the mechanisms that prevent or sometimes favor their release in natural waters.

The term *binding of pesticides* in this chapter refers to chemical reactions of pesticide residues with both low molecular, water-soluble humic fractions and highly polymerized, insoluble fractions. In the latter case, the International Union of Pure and Applied Chemistry (IUPAC) definition of soil-bound pesticide residues applies. According to this definition, soil-bound residues are "chemical species originating from pesticides used according to good agricultural practices that are unextracted by methods which do not significantly change the chemical nature of the residues."[6]

According to information available today, the chemical classes of phenols and anilines, as well as those pesticides that can generate these substances as metabolites, have the highest tendency to be bound to humic substances.[7,8] Therefore, chlorinated phenols and anilines will be the preferred model compounds in this chapter. Contrarily, chlorinated hydrocarbons in general have the lowest tendency to form bonds with humic substances,[7,8] therfore the use of some representatives of this class as model pesticides is of interest. Another class of pesticides for which binding capacity to humus is well investigated and will also be discussed is that of s-triazines.[9-14]

The chapter will consider successively the presence and behavior of pesticides in soil leachates, the interactions of these compounds with soil humus components, the biodegradability, and finally the transfer of pesticides and their metabolites to plants.

II. PESTICIDES IN SOIL LEACHATES

A. The Presence of Pesticides in Natural Waters

The solubility of pesticides and related compounds in water is as diverse as their chemical structures, and commonly ranges between a few milligrams per liter to more than 1 kg/L as in the case of vamidothion (4 kg/L). However, migration processes in soils are not only dependent on the solubility of the molecules, but also on other physicochemical properties (i.e., ionizability and reactivity with mineral and organic surface groups, including hydrophobic binding) and their persistence in soil media (i.e., volatility and resistance to biodegradation).[15] Pesticide molecules can be detected in natural waters, either in their unchanged active form or as metabolites and products of chemical reactions with soluble humus, or both. The concentration of pesticides in natural waters can even be higher than their theoretical water solubility; possibly this is due to the presence of dissolved organic substances, especially humic molecules, which

Table 1. Bound and Total Residues of [¹⁴C]Aldrin in Soil
for 10 Years After Application

Year	Bound residues[a]	Total residues[a]
1969, After application	0	2.72
1969, After harvest	0.17	1.87
1970	0.11	1.18
1971	0.06	0.57
1972	0.13	0.64
1973	0.11	0.50
1974	0.09	0.44
1975	0.11	0.43
1979	0.11	0.35

Source: Klein and Scheunert.[7]

[a] μg (Equivalent to aldrin)/g of soil.

were shown to enhance the solubility of pesticides.[16] For example, the concentration of substituted anilines, especially the chlorinated ones, which participate in the structure of numerous herbicides and a few fungicides, can reach 30 μg/L in European streams. Such contamination is mainly due to chemical industries.[17] In the case of atrazine, annual losses estimated by different authors range between 0.2 and 2% of the applied doses for runoff and between 0.005 and 2% for leaching. Reported concentrations of atrazine in drainage waters range between 6 and 300 μg/L.[18] In order to face the risks of contamination of drinking water, the European Communities have issued a directive which would limit the maximal acceptable concentration of pesticides to 0.1 μg/L for each individual pesticide, and 0.5 μg/L for the sum of all pesticides present.[19]

B. Lysimeter Experiments

Any organic compound, once added to the soil, distributes into three main fractions: solvent-extractable, water-extractable, and humus-bound fractions, for which distribution changes with time at a rate that depends on the factors mentioned above.[20] This also applies in the case of pesticides and other xenotiotics.[6] Although the total amount of residues stored in the first few centimeters of soil decreases with time, the ratio of bound to extractable residues increases as a result of their structural modification,[7] as observed in the case of aldrin (1,2,3,4,10,10-hexachloro-1,4,4a,5,8,8a-hexahydro-exo-1,4-endo-5,8-dimethanonaphthalene). ¹⁴C-labeled aldrin was applied at a rate of 2.9 kg/ha to outdoor soil having the following composition: sand 67.3%, silt 16.7%, clay 12.5%, organic matter 3.5%, and pH 8.1. By extraction with methanol in a Soxhlet, extractable ¹⁴C was removed; unextractable ¹⁴C left in soil after this procedure was considered as *soil-bound*. Total as well as soil-bound residues are listed in Table 1. Such a binding can be reversible: discrete amounts of water-soluble material can be released and translocated either laterally or vertically through soil pores.

Lysimetric field devices are designed to collect leachates and soil samples, where the translocated products are then quantified and identified.[6,11,12] Nuclear methods, mainly with [14]C- and ^{15}N-labeled pesticide molecules, are used to track and localize them among the soil layers and components. Standardized leaching experiments are frequently designed for studying at the same time the degradability and mobility of pesticides in soils, at least during one seasonal cycle.[21,22] These methods are of special interest for the purpose of registration of new active molecules.[23] Furthermore, leaching experiments aim at quantifying the risks of contamination of groundwaters and identifying the chemical forms of the migrating compounds. In this case, outdoor experiments can be run for several years.

Scheunert et al.[24] conducted assays in $60 \times 60 \times 70$ cm wooden boxes, which lay in pits of similar size, and were filled with about 160 kg of the same soil as in the surrounding field. In the bottom of each box, a grid allows the draining waters to pass through and be collected on a metallic plate. In such a device, these authors studied the fate of several ^{14}C-labeled compounds (i.e., aldrin, buturon, 4-chloroaniline) for more than 15 years. The concentrations of the compounds recovered in the effluents (expressed on the basis of their radioactivity) increased during the first year, passed through a maximum after 1 to 3 years, and then decreased at a rate which depended on the chemical structure of the pesticide.

In the case of the herbicide buturon (N-[4-chlorophenyl]-N'-methyl-N'-isobutinyl-urea), two approaches were used in which the molecule was labeled either on the benzene ring or on the N'-methyl group. The soil used in this case was a sandy soil with the following characteristics: sand 76.2%, parasilt 10.3%, clay 13.5%, organic matter 1.2%, and pH 7.0. Initial loading rate was 3 kg/ha in two successive years in the experiment with ring-labeled compounds, and 2 kg/ha in the first year in the experiment with N'-methyl-labeled compounds. After 12 years, the radioactivity collected in water amounted to 2.14 and 1.66% of total ^{14}C applied, in the respective approach[25] (Figure 1). Gas chromatography (GC) and mass spectrometry (MS) analysis of the compounds extracted from the leachates of the ^{14}C ring-labeled molecules with dichloromethane indicated that 4-chloroaniline was the main conversion product, and represented about 16% of the radioactivity (or 53 ng/L) in the leachates.

Numerous field experiments and measurements have been conducted in the case of another herbicide, atrazine (6-chloro-N-ethyl-N'-[1-methylethyl]-1,3,5-triazine-2,4-diamine), with a water solubility of 0.028 g/L at 20°C and an octanol/water partition coefficient (K_{ow}) of 440.8. Schiavon[11] added ^{14}C ring-labeled atrazine at an initial rate of 1.6 kg/ha to a cultivated *sol brun lessivé* which had the following characteristics: clay 30%, organic carbon 1.6%, pH 6.5, and cation exchange capacity 0.15 meq/g. With this experiment he showed that the leached quantities depended on the amounts of herbicide applied, and on the soil adsorbing capacity and porosity: at the end of the summer 4 months after the start

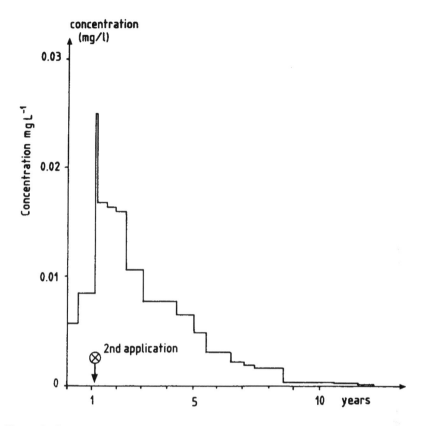

Figure 1. Concentration of radioactive compounds in leachates (depth 60 cm) after application of [¹⁴C]buturon on soil surface. (After Reiml, D., I. Scheunert, and F. Korte. *J. Agric. Food Chem.* 37:244–248 [1989]. With permission.)

of the experiment, the sum of radioactivity measured in the leachates was less than 0.23% of the total applied, then reached about 4% in the winter period, and amounted to 5–6% after 13 months. Although the sum of total leached radiocarbon increased, the concentration of leached ¹⁴C atrazine in the water decreased after this time. Figure 2 shows that unidentified water-soluble compounds predominated only at the very beginning of the experiment: most of the atrazine and its metabolites were transported in their free form in soil water, and dealkylated forms predominated throughout the experiment.

The use of labeled molecules in the above examples allows us to study simultaneously the transfer of the active substance and of its main metabolites. In natural conditions and in the absence of such labeled material, a strong analytical effort is needed to determine the actual pollution level of soils and groundwaters. Unidentified compounds probably result from transient reactions between pesticide degradation products and natural solutes, and can also account

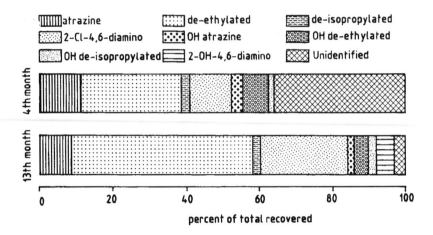

Figure 2. Composition of percolation water from soil column treated with [¹⁴C]-atrazine, at two different periods following treatment. (Adapted from Schiavon, M. *Ecotoxicol. Environ. Saf.* 15:46–54 [1988].)

for pollution. In the case of atrazine, more accurate methods are required to determine chlorinated and hydroxylated metabolites.[18]

III. PESTICIDES IN HUMIC SUBSTANCES

The decreased leaching out of pesticides and the increasing proportion of their bound residues are generally indicative of their increasing incorporation into water-soluble humic substances.[11,12] Schematic structures were early proposed by Stevenson[26] in the case of cationic pesticides (diaquat, s-triazines) to illustrate their reactions with humic molecules. Again in the case of atrazine, Capriel et al.[27] applied this ¹⁴C-labeled herbicide to a mineral soil under field conditions. After 9 years, 50% of the ¹⁴C residues were found in the bound (i.e., unextractable with methanol in a Soxhlet) form. These bound ¹⁴C residues were distributed among the various soil humic fractions. In addition to the parent herbicide, a considerable proportion (about 75%) of these residues was hydroxy analogues of atrazine and their dealkylated products.

Bertin[12] has conducted an experiment under outdoor conditions, in which ¹⁴C ring-labeled atrazine was applied to the top of soil columns filled with three different soil types. These were a *pelosol* (clay 52%, organic carbon 1.3%, pH 7.3), a *sol brun lessivé* (clay 24%, organic carbon 1.6%, pH 6.4), and a *rendzina* (clay 20%, organic carbon 10;5%, pH 7.4). The application rate corresponded to 2 kg/ha. The solutions percolated through the columns were collected, and the distribution of the remaining radioactivity in the soil grain-size fractions and in the alkali-soluble humus was studied. At the beginning of the experiment, the radioactivity not extractable with methanol was mainly associated to the low

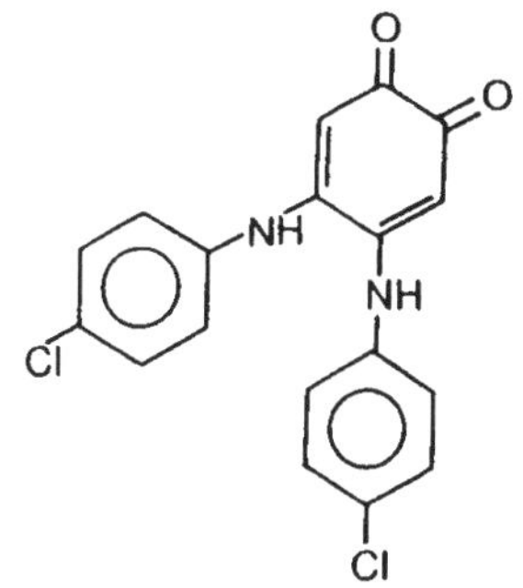

Figure 3. Identified structure of the main addition product of 4-chloroaniline and catechol. (Adapted from Adiran, P., E.S. Lahaniatis, F. Andreux, M. Mansour, I. Scheunert, and F. Korte. *Chemosphere* 7–8:1599–1609 [1989].)

molecular size fraction (i.e., fulvic acids) and then decreased with time, whereas that present in the humic acids and in the extractable humin fractions progressively increased. The distribution was different from one soil to another, but a tendency toward a higher incorporation into humic acids and the humin in soils with higher organic carbon content was noticed. However, little is still known about the nature of the reacting compounds.[28,29]

The nature of pesticide-bound residues has been established in some cases, e.g., substituted anilines.[17,30-32] These compounds are readily reactive under oxidizing conditions and give conjugated products with naturally occurring phenols which can participate in the structure of polymeric molecules. During the oxidative polymerization of o- and p-diphenols, oxygen transfer reactions are generally controlled by phenolase exoenzymes of plant and microbial origins, although metal-catalyzed reactions are involved. Metallic catalysts, especially divalent Fe and Mn, not only stimulate polyphenol oxidation, but also enhance the incorporation of amino compounds into water-soluble, humic-like polycondensates.[32,33]

By oxidizing phenolic precursors of humic-like polymers in the presence of 4-chloroaniline, quinonic building blocks were obtained (Figure 3), and experimental conditions were tested in order to optimize the formation of these products. It could be established that the incorporation of 4-chloroaniline into anilinoquinone and catechol-deriving polymeric structures increased in the presence of Mn, especially Mn dioxide (pyrolusite), as shown in Figure 4. Polymeric-bound residues formed in these conditions are generally water soluble, but they easily form insoluble "complexes" with metal ions and clay mineral surfaces.[17,25]

The incorporation of other pesticide molecules into phenolic polymers was attempted, especially in the case of methoxychlor (1,1'-[2,2,2-trichloroethylidene]bis[4-methoxybenzene]),[34] and atrazine and its metabolites.[35] In the latter case, structure of the products could not be established; however, it was clearly shown that dealkylated metabolites of atrazine yielded much higher amounts of

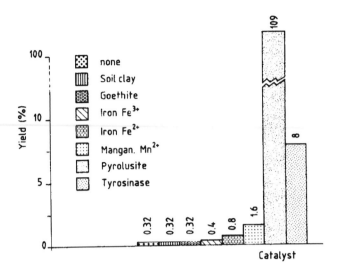

Figure 4. Yields of anilino-quinone in the presence of different catalysts. (Adapted from Adrian, P., E.S. Lahaniatis, F. Andreux, M. Mansour, I. Scheunert, and F. Korte. *Chemosphere* 7–8:1599–1609 [1989].)

polymers, probably due to the formation of covalent bonds through the $-NH_2$ groups of these metabolites.[36]

The binding of pesticides to humic molecules through photochemical processes has been reported, at least from laboratory experiments. Choudry[37] obtained a low yield of benzofurane derivatives from the photolysis of a solution mixture of tetrachlorobenzene and phenol. The photoincorporation of pesticides into polymeric compounds is therefore possible, provided a quinonic intermediate can appear.[38] Research about these reaction mechanisms needs to be developed because numerous reactive degradation products of pesticides can be obtained through photosensitive reactions.[39]

IV. BIOTRANSFORMATIONS AND BIOAVAILABILITY OF PESTICIDES

Pesticides can be degraded, transformed, or stored by microorganisms, plants, and animals. They can undergo numerous structural changes, and their metabolites can exhibit different toxic effects as compared to the parent compounds.

A. Effects of Pesticides on Soil Metabolic Processes

It is well known that pesticides affect the metabolic processes of soil microorganisms, as well as plant pathogens,[40] to variable extents. Although this aspect is distant from the purpose of the present chapter, it is important to keep in mind that pesticide molecules or their metabolites can modify the biodegradation of soil organic matter through selective effects on soil enzymic processes. This is well known in the case of nitrification, which can be strongly inhibited by herbicides.[41,42] In the same way, it can be expected that the lethal effect of herbicides on the growth of soil fungi[43] could affect decomposition processes. Thus, decreases in soil fertility by the accumulation of raw humus on the soil surface may occur.

B. Degradation of Soil-Bound Pesticide Residues by Soil Microorganisms

The way bound residues are degraded in soils and the nature of the organisms involved still require studies. Fungal degradation of humic-bound 3,4-dichloroaniline was studied by Hsu and Bartha.[44] These authors showed that in culture medium two fungi (*Penicillium frequentans* and *Aspergillus versicolor*) were able to degrade humic acids, as well as humic-bound 3,4-dichloroaniline. In a first degradation pathway, the mineralization of the aniline ring occurred, either after release of the aniline moiety or when it was still attached. The second pathway was the release of soluble oligomers with intact bound dichloroaniline moieties. This suggests that these fragments of pesticide residues, if added to growing plants, could be bioavailable to these plants.[45]

More recently, another conclusion regarding the biodegradability of bound residues in soils was drawn by Dec and Bollag,[46] who showed that the release of chlorophenol bound to synthetic humic acids prepared by oxidative polymerization of phenolic molecules was very slow under experimental conditions. The amount of released compounds which were converted into CO_2 did not exceed 10% of the initial bound material after more than 10 weeks, and the rest remained completely associated with the polymers.

By reacting 3,4-dichloroaniline with catechol, Völkel et al.[47] obtained a similar quinonic reaction product as in the case of 4-chloroaniline (Figure 3). When incubated in an agricultural soil, this product was shown to be converted into CO_2 with a 10-fold slower rate than the corresponding free 3,4-dichloroaniline molecule.

Apart from synthetic bound residues, it is also possible to compare the release of pesticide derivatives bound to different soil fractions under defined incubation conditions. In the examples shown in Tables 2 and 3, the behavior of ^{14}C-labeled 4-chloroaniline and 2,4,6-trichloroaniline added to a soil was studied.[48] Soil composition was as follows: clay 33.6%, silt 27.4%, sand 32.4%, coarse matter 6.6%, organic matter 3.15% and pH 6.4. The labeled contaminants were applied at the respective concentrations 2.28 and 1.79 mg/kg. Extraction of nonbound ^{14}C was performed with methanol in a Soxhlet, next with acetate buffer solution (pH 4.6), and then with methanol in a Soxhlet again. Aerobic biomineralization

Table 2. Concentrations of Soil-Bound Residues of [^{14}C] 4-Chloroaniline in Soil Fractions, and Their Mineralization and Photodegradation

Soil fraction	Conc. of bound ^{14}C (μg/g^a)	Aerobic mineralization 28 days (% ^{14}CO$_2$)	Photodegradation $\lambda > 290$ nm 17 hr (% ^{14}CO$_2$)
Inorganic fractions	0.526	1.73	0.59
Organic fractions			
Humic acids	30.7	1.21	0.08
Humin	6.97	4.13	1.18
Fulvic acids	5.95	4.98	1.43
Total soil	1.50	2.18	1.39
Free 4-chloroaniline	—	9.04	26.06

Source: Ter Meer-Bekk.[48]

^{a} μg (Equivalent to the parent compound)/g dry soil fraction.

and abiotic photodegradation of 4-chloroaniline and 2,4,6-trichloroaniline bound to various soil fractions were considerably slower than those of the respective free compounds. The compounds bound to the fulvic acid fraction were more degraded than those bound to the humic acid fraction, probably due to the difference in the polymerization degree of these fractions. Aerobic and anaerobic biomineralization of 2,4,6-trichloroaniline was an exception: it was higher for bound residues than for the free compound. On the basis of composition of the nonbound material, it was suggested that 2,4,6-trichloroaniline was probably not bound as an unchanged compound, but as metabolites (such as dichlorinated products) which were more easily biodegraded.

C. The Bioavailability of Pesticides to Growing Plants

1. Incorporation Processes

Bound residues can be bioavailable to growing plants as well as to animals and become a source of contaminants, depending on the degree of their transformation inside the plant or animal. Three processes can generally be distinguished: the first one is governed by biological reactions, such as enzymic oxidation, reduction, hydrolysis, or acylation. In the second one, the pesticide or its metabolite form conjugation derivatives with simple biomolecules. In the plants and animals, important effort was made to identify these conjugates.[49] In the third case, the conjugation occurs with reactive functional groups of more complex biopolymers, either in the vacuoles with humic-like polymers or with lignin.[50]

The reversibility of these reactions may be possible only in the first and second cases, but it is very unlikely and slow in the third case. These transformations that occur in living organisms offer the possibility for changing the chemical

Table 3. Concentrations of Soil-Bound Residues of [^{14}C]2,4,6-Trichloroaniline in Soil Fractions, and Their Mineralization and Photodegradation

Soil fraction	Conc. of bound ^{14}C (μg/g^a)	Aerobic mineralization 28 days (% ^{14}CO$_2$)	Photodegradation $\lambda = >290$ nm 17 hr (% ^{14}CO$_2$)
Inorganic fractions	0.019	1.08	0.77
Organic fractions			
Humic acids	1.03	0.39	0.09
Humin	0.24	1.39	1.11
Fulvic acids	0.22	5.05	1.48
Total soil	0.05	1.91	1.44
Free 2,4,6-trichloroaniline	—	0.614	29.88

Source: Ter Meer-Bekk.[48]

^{a} μg (Equivalent to the parent compound)/g dry soil fraction.

nature of normally poorly water-soluble xenobiotics and for incorporating them into strongly polar compounds, which can be excreted by animals or released by plants through exudation and decay processes. Because of their bioavailability, these conjugated pesticides are considered as potentially toxic.[51-53] The terminal reaction products which are covalently bound to high molecular weight polymers are more stable chemical entities than other conjugates. Due to their insoluble character, which diminishes the possibilities of enzymic attack, they are of low bioavailiability and are considered of low toxicological interest.

The role of soil fauna on the fate of xenobiotics in soil needs to be better integrated in this kind of study. It is well known that the presence of earthworms stimulates soil microbial activity,[54] and therefore can influence indirectly the enzymic transformation of extraneous molecules.[55] Increased studies about bioaccumulation of pesticides by the soil fauna, in the same way as for plants, would contribute to the finding of new indicators of pollution or risks of pollution and to understand (and perhaps forecast) the processes of long-term release of toxic substances from bound residues.

2. Measurements

In the case of plants in the field as well as in pot experiments, difficulties in the measurement of the bioavailability of a pesticide or related compound present in the soil arise from assessment of the proportion of soil volume which is effectively explored by plant roots. This problem is especially met in studies with herbicides, when different plants or different soil types are compared. In most cases ^{14}C-labeled molecules are used, and the uptake by plants is determined as a percentage of the total radioactivity added to the soil-plant systems.

These percentages rarely exceed 2.5%, with larger proportions in the roots than in the aerial parts.[8] In the case of dinitroanilines (butralin, chlornidin,

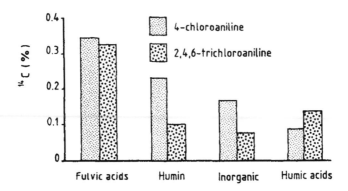

Figure 5. Uptake of residues of ¹⁴C-labeled chlorinated anilines by barley seedlings (in % of ¹⁴C in the respective fractions). (After ter Meer-Bekk, C. Doctor Thesis, Technische Universität München, Germany [1986]. With permission.)

dinitramin, fluchloralin, profloralin, trifluralin), the maxima obtained by Helling et al.[56] was 1.07% in soybean after 10 weeks. Khan[50] obtained 0.53% for prometryn (N,N'-bis-[1-methylethyl]-6-[methylthio]-1,3,5-triazine-2,4-diamine) in oats after 3 weeks. In an experiment with triazine ring-labeled atrazine incorporated in natural and synthetic humic acids, Bertin[12] showed that 0.7 and 1.7% of the initial radiocarbon were taken up by maize plants, respectively. In the plant, atrazine was mostly in a conjugated form, although the presence of hydroxylated derivatives could not be unequivocally demonstrated.

The bioavailiability of residues of a given molecule can be very different according to the nature of the soil fraction they are bound to, as shown by Ter Meer-Bekk[48] in the case of the uptake by barley seedlings of 4-chloroaniline and 2,4,6-trichloroaniline. In this experiment, soil composition, application rates, and extraction of nonbound residues were the same as discussed above for aerobic biomineralization experiments. Besides, this author established (Figure 5) that bound residues in fulvic acids were clearly more bioavailable than those in humic acids and inorganic fractions, probably in relation with the mobility and polymerization degree of the respective fractions. Nevertheless, the uptake of free anilines by plants considerably exceeded that of anilines bound in every soil fraction.

V. SUMMARY AND CONCLUSIONS

By the interaction of pesticides and/or their primary metabolites with both water-soluble and insoluble humic substances, new xenobiotic derivatives are formed, for which the chemical structure is widely unknown so far. Binding of pesticide residues to water-soluble humic fractions can enhance leaching to groundwater, even for those pesticides which normally have a very low water solubility. Binding to highly polymerized unsoluble humic fractions greatly reduces the mobility of pesticides in soil bioavailability. There are numerous experimental evidences that bound residues can pass from the water-soluble to the solid humic phase and vice versa. However, information available today is not sufficient to forecast these transfers in natural conditions and to evaluate their specific environmental and ecotoxicological consequences. The binding of pesticide residues to natural organic matter can be considered a key process that can be involved in the contamination of groundwater as much as in detoxification processes, depending on both the chemical nature of the pesticide and the characteristics of this organic matter.

REFERENCES

1. Scheffer, F. and P. Schachtschabel. *Lehrbuch der Bodenkunde,* 9th ed. in German (Stuttgart: Ferdinand Enke Verlag, 1976).
2. Duchaufour, P. *Pédologie, Vol. 1. Pédogénèse et Classification* (Paris: Masson, 1977), pp. 1–167.
3. Schnitzer, M. and P.M. Huang, Eds. *Interactions of Soil Minerals with Natural Organic and Microbes,* S.S.A. Special Publication number 17 (Madison, WI: Soil Science Society of America, 1986).
4. Sawhney, B.L. and K. Brown. Eds. *Reactions and Movements of Organic Chemicals in Soils,* S.S.S.A. Special Publication number 22 (Madison, WI: Soil Science Society of America, 1989).
5. Korte, F. "Concepts for the Ecotoxicological Evaluation of Chemicals: Ecotoxicological Profile Analysis," in *Pollutants and Their Ecotoxicological Significance,* H.W. Nürnberg, Ed. (John Wiley & Sons, 1985), pp. 337–361.
6. Führ, F. "Non-Extractable Pesticide Residues in Soil," in *Pesticide Science and Biotechnology,* R. Greenhalgh and T.R. Roberts, Eds. (Oxford: International Union of Pure and Applied Chemistry, Blackwell Scientific Publications, 1987), pp. 381–389.
7. Klein, W. and I. Scheunert. "Bound Pesticide Residues in Soil, Plants and Food with Particular Emphasis on the Application of Nuclear Techniques," in *Agrochemicals: Fate in Food and the Environment* (Vienna: IAEA-SM-263/38, International Atomic Energy Agency, 1982), pp. 177–205.
8. Roberts, T.R. "Non-extractable Pesticide Residues in Soils and Plants," *Pure Appl. Chem.* 56:945–956 (1984).
9. Li, G.C. and G.T. Felbeck, Jr. "A Study of the Mechanism of Atrazine Adsorption by Humic Acid from Muck Soil," *Soil Sci.* 113:140–148 (1972).
10. Senesi, N. and C. Testini. "Theoretical Aspects and Experimental Evidence of the Capacity of Humic Substances to Bind Herbicides by Charge-Transfer Mechanisms (Electron-Donor-Acceptor Processes)," *Chemosphere* 13:461–468 (1984).
11. Schiavon, M. "Studies of the Leaching of Atrazine, of Its Chlorinated Derivatives, and of Hydroxyatrazine from Soil Using ^{14}C Ring-Labeled Compounds Under Outdoor Conditions," *Ecotoxicol. Environ. Saf.* 15:46–54 (1988).

12. Bertin, G. "L'Immobilisation de l'Atrazine par la Matière Organique des sols. Une Approche Modélisée en Conditions Naturelles et au Laboratoire," Doctor Thesis, Institut National Polytechnique de Lorraine, Nancy, France (1989).

13. Bertin, G. and M. Schiavon. "Les Résidus non Extractibles de Produits Phytosanitaires dans les Sols," Agronomie 9:117–124 (1989).

14. Winkelmann, D.A. and S.J. Klaine. "Degradation and Bound Residue Formation of Atrazine in a Western Tennessee Soil", Environ. Toxicol. Chem. 10:335–345 (1991).

15. Weber, J.B. and C.T. Miller. "Organic Chemical Movement over and through Soil," in Reactions and Movements of Organic Chemicals in Soils, B.L. Sawhney and K. Brown, Eds. (Madison, WI: Soil Science Society of Agronomy, 1989), pp. 305–334.

16. Chiou, C.T., R.L. Malcolm, T.I. Brinton, and D.E. Kile. "Water Solubility Enhancement of Some Organic Pollutants and Pesticides by Dissolved Humic and Fulvic Acids," Environ. Sci. Technol. 20:502–508 (1986).

17. Adrian, P., F. Andreux, R. Viswanathan, D. Freitag, and I. Scheunert. "Fate of Anilines and Related Compounds in the Environment. A review," Toxicol. Environ. Chem. 20–30:109–120 (1989).

18. Schiavon, M., J.M. Portal, and F. Andreux. "Données Actuelles sur les Transferts d'Atrazine dans l'Environnement," Agronomie 12:129–139 (1992).

19. Commission of the European Communities. Directive "Qualité de l'Eau Destinée à la Consommation Humaine," n° 80/778/CEE, Paramètre 55.

20. Carballas, T., F. Andreux, and F. Jacquin. "Répartition des Principaux Constituants d'un Végétal Marqué au ^{14}C dans les Composés Humiques d'un Sol à Mull," Sci. Sol. (Paris) 3:29–38 (1971).

21. Brumhard, B., F. Führ, and W. Mittelstaedt. "Leaching Behaviour of Aged Pesticides: Standardized Soil Column Experiments with ^{14}C-Metamitron and ^{14}C-Methabenzthiazuron," in British Crop Protection Conference-Weed 2, (1987), 585–592.

22. Jamet, P. "Etude Simultanée de la Dégradation et de la Mobilité d'un Pesticide et de ses Principaux Métabolites dans le sol," in Symposium I.N.R.A.-F.A.O. Comportement et Effets Secondaries des Pesticides dans le Sol, (Versailles: Service des Publications de l'I.N.R.A., 1985) 163–171.

23. Biologische Bundesanstalt für Land- und Forstwirtschaft, Bundesrepublik Deutschland. Richlininien für die Prüfung von Pflanzenschutzmitteln im Zulassungverfahren, Teil IV:4–3 (1990).

24. Scheunert, I., F. Korte, and D. Reiml. "Applikationsuntersuchungen von Pflanzenbehandlungmitteln im System Pflanze/Boden in Lysimetern unter Einsatz ^{14}C-markierter Substanzen und der besonderen Berücksichtigung des Leaching-Verhaltens," in Schr. Reihe Verein WaBoLu 68 (Stuttgart: Gustav Fischer Verlag, 1987) pp. 313–322.

25. Reiml, D., I. Scheunert, and F. Korte. "Leaching of Conversion Products of (^{14}C) Buturon from Soil during 12 Years after Application," J. Agric. Food Chem. 37:244–248 (1989).

26. Stevenson, F.J. "Role and Function of Humus in Soil with Emphasis on Adsorption of Herbicides and Chelation of Micronutrients," Bioscience 22(11):643–650 (1972).

27. Capriel, P., A. Haisch, and S.U. Khan. "Distribution and Nature of Bound (Non-extractable) Residues of Atrazine in a Mineral Soil Nine Years after the Herbicide Application," *J. Agric. Food Chem.* 33:567–569 (1985).

28. Andreux, F., M. Schiavon, G. Bertin, J.M. Portal, and E. Barriuso. "The Use-fulness of Humus Fractionation Methods in Studies About the Behaviour of Pollutants in Soils," *Toxicol. Environ. Chem.* 31–32:29–38 (1991).

29. Barriuso, E., M. Schiavon, F. Andreux, and J.M. Portal. "Localization of Atrazine Non-Extractable (Bound) Residues in Soil Size Fractions," *Chemosphere* 22, 12:1131–1140 (1991).

30. Bollag, J.M., R.D. Minard, and S.-Y. Liu. "Cross-Linkage Between Anilines and Phenolic Humus Constituents," *Environ. Sci. Technol.* 17:72–80 (1983).

31. Simmons, K.E., R.D. Minard, and J.M. Bollag. "Oxidative Co-Oligomerization of Guaiacol and 4-Chloroaniline," *Environ. Sci. Technol.* 23:115–120 (1989).

32. Adrian, P., E.S. Lahaniatis, F. Andreux, M. Mansour, I. Scheunert, and F. Korte. "Reaction of the Soil Pollutant 4-Chloroaniline with the Humic Acid Monomer Catechol," *Chemosphere* 7–8:1599–1609 (1989).

33. Adrian, P., F. Andreux, M. Metche, M. Mansour, and F. Korte. "Autoxydation des Ortho-Diphenols Catalysée par les Ions Fe²⁺ et Mn²⁺: un Modèle de Formation des Acides Humiques," *C. R. Acad. Sci. Paris* 303II(17):1615–1618 (1986).

34. Mathur, S.P. and H.V. Morley. "Incorporation of Methoxychlor-¹⁴C in Model Humic Acids Prepared from Hydroquinone," *Bull. Environ. Contam. Toxicol.* 20:268–274 (1978).

35. Bertin, G., M. Schiavon, J.M. Portal, and F. Andreux. "Contribution to the Study of Non-Extractable Pesticide Residues in Soils: Incorporation of Atrazine in Model Humic Acids prepared from Catechol," in *Diversity of Environmental Biogeochemistry*, J. Berthelin, Ed. (Amsterdam, Netherlands: Elsevier, 1991), pp. 105–110.

36. Andreux, F., J.M. Portal, M. Schiavon, and G. Bertin. "The Binding of Atrazine and Its Dealkylated Derivatives to Humic-Like Polymers Derived from Catechol," *The Science of the Total Environment* 117/118:207–217 (1992).

37. Choudry, G.C. "Humic Substances: Structural Aspects and Photophysical, Photochemical and Free Radical Characteristics," in *The Handbook of Environmental Chemistry, Vol. 1*, Part C, O. Hutzinger, Ed., (Berlin, West Germany: Springer-Verlag, 1984), pp. 1–24.

38. Wang, W.H., R. Beyerle-Pfnür, and J.P. Lay. "Photoreaction of Salicylic Acid in Aquatic Systems," *Chemosphere* 17(6): 1197–1204 (1988).

39. Mansour, M., E. Feicht, and P. Méallier. "Improvement of the Photostability of Selected Substances in Aqueous Medium," *Toxicol. Environ. Chem.* 20–21:139–147 (1989).

40. Altman, J. and C.L. Campbell. "Effect of Herbicides on Plant Diseases," *Annu. Rev. Pythopathol.* 15:361–383 (1977).

41. Gaur, A.C. and K.C. Misra. "Effect of Simazine, Lindane, and Ceresan on Soil Respiration and Nitrification Rates," *Plant Soil* 46:5–15 (1977).

42. Dubey, H.D. "Effect of Picloram, Diuron, Ametryne, and Prometryne on Nitrification in Some Tropical Soils," *Soil Sci. Soc. Am. Proc.*, 33:893–896 (1969).

43. Isakeit, T. and J.L. Lockwood. "Lethal Effect of Atrazine and Other Triazine Herbicides on Ungerminated Conidia of *Cochliobolus sativus* in Soil," *Soil Biol. Biochem.* 21(6):809–817 (1989).

44. Hsu, T.S. and R. Bartha. "Biodegradation of Chloroaniline-Humus Complexes in Soil and in Culture Solution," *Soil Sci.* 118(3):213–220 (1974).

45. Still, C.C., T.S. Hsu, and R. Bartha. "Soil Bound 3-4 Dichloroaniline: Source of Contamination in Rice Grain" *Bull. Environ. Contam. Toxicol.* 24:550–554 (1980).

46. Dec, J. and J.M. Bollag. "Microbial Release and Degradation of Catechol and Chlorophenols Bound to Synthetic Humic Acid," *Soil Sci. Soc. Am. J.* 52(5):1366–1371 (1988).

47. Völkel, W., T. Choné, M. Mansour, and F. Andreux. "Dégradation dans un Sol Brun de Culture, de la 3,4-Dichloroaniline Libre ou Liée à un Monomère de Type Humique," in Réactions et Biotransformations des Xénobiotiques, 21ème Congrès du Groupe Français des Pesticides Nancy-Brabois, (1992) 355–359.

48. Ter Meer-Bekk, C. "Bildung, Charakterisierung und Bedeutung sogenannter "Gebundener Rückstände" in Boden," Doctor Thesis, Technische Universität München, Germany (1986).

49. Paulson, G.D., J.C. Caldwell, D.H. Hutson, and J.J. Menn. Eds. *Xenobiotic Conjugation Chemistry,* A.C.S. Symposium Series 299 (Washignton, DC: American Chemical Society, 1986).

50. Khan, S.U. "Plant Uptake of Unextracted (Bound) Residues from an Organic Soil Treated with Prometryn," *J. Agric. Food Chem.* 28:1096–1098 (1980).

51. Burgat-Sacaze, V., A.G. Rico, and J.C. Panisset. "Toxicological Significance of Bound Residues," in *Drug Residues in Animals,* A.G. Rico, Ed. (New York: Academic Press, 1986), pp. 1–31.

52. Mikami, N., J. Yushimura, H. Kaneko, H. Yamada, and J. Miyamoto. "Metabolism in Rats of 3-Phenoxybenzyl Alcohol and 3-Phenoxy-Benzoic Acid Glucoside Conjugates Formed in Plants," *Pest. Sci.* 16:33–45 (1985).

53. Mulder, G.J., J.H. Meerman, and A.M. van den Goorbergh. "Bioactivation of Xenobiotics by Conjugation," in *Xenobiotic Conjugation Chemistry,* G.D. Paulson, J.C. Caldwell, D.H. Huston, and J.J. Menn, Eds. (Washington, DC: American Chemical Society, 1986), pp. 282–300.

54. Toutain, F. "Les Phénomènes de Biodégradation et d'Humification dans les Ecosystèmes Forestiers Tropicaux: Rôle et Conséquences de la Diversité Spécifique," in Report on Meeting of I.U.B.S. Working Group on Species Diversity/Decade of the Tropics. Programme "The Significance of Species Diversity in Tropical Forest Ecosystems," *Biol. Int.* Special Issue, 6:19–29 (1984).

55. Reinecke, A. and R.G. Nash. "Toxicity of 2,3,7,8-TCDD and Short-Term Bioaccumulation by Earthworms (Oligocheata)," *Soil Biol. Biochem.* 16(1):45–49 (1984).

56. Helling, C.S. and A.E. Krivonak. "Biological Characteristics of Bound Dinitroaniline Herbicides in Soils," *J. Agric. Food Chem.* 26:1164–1172 (1978).

CHAPTER 14

Conversion of Lindane to HCH Isomers and HCB in the Agricultural Field Conditions

Stefan M. Waliszewski

I. INTRODUCTION

Lindane, γ isomer of hexachlorocyclohexane (HCH), has been used for decades to control numerous species of insects in field crops.[1,9,23] The technical product of HCH (which contains: 60–70% of α-HCH, 5–12% β-HCH, 10–15% γ-HCH, 6–10% δ-HCH, and 3–4% ϵ-HCH[23] in many tropical and subtropical countries) is used and recommended in agriculture and sanitation, at the dose of 2.5 kg of active ingredient per hectare.[2,4,11,16] The use of technical HCH has been banned in western countries, but it still finds extensive use in several developing and tropical countries as the insecticide of choice in the plant protection and vector control programs.[2,11,15] For example, in India alone the current consumption of technical HCH is approximately 36 million kilograms per year.[15]

In the 1980s, an extensive discussion among agricultural ecologists took place to explain the residue sources of HCH isomers in the soil, plant products, and milk. Despite the fact that many years ago technical HCH was banned worldwide and only a few governments permitted the use of lindane in agriculture, it still was observed in different elements of the environment and food chain residues of α- and β-HCH. Lindane was blamed for this state of affairs and was assumed to bioisomerize in the agricultural environment to α-, β-, and δ-HCH.[3,6,7,10,12-14,17-25] This bioisomerization phenomenon of lindane was studied in laboratory conditions which permitted the forming of artificial environmental conditions never met in nature. Field conditions have greater variability that

0-87371-616-7/93/$0.00 + $.50

Table 1. Contents (mg/kg) of Lindane,
HCH Isomers, and HCB in the
Applied Preparation "Owadziak
Pylisty 2.4"

Compounds	First year	Second year
Lindane	20.0206	20.9624
α-HCH	0.0510	0.0509
β-HCH	n.d.[a]	n.d.
δ-HCH	0.0156	0.0185
ε-HCH	n.d.	n.d.
HCB	n.d.	n.d.

[a] Not detected.

cannot be controlled. For this reason, field investigations for the bioisomerization phenomenon of lindane were prompted to evaluate the magnitude and confirm previous laboratory results. In this investigation two primary objectives were taken into account:

1. determination of the loss of lindane and HCH isomers formed from lindane in the soil after lindane application in the dose of 1.7 kg of active ingredient per hectare, recommended by the Institute of Plant Protection in Poznań, Poland for good agricultural practice[29]
2. evaluation of the magnitude of bioisomerization phenomena of lindane to α-, β-, δ-, ε-HCH and HCB in the soil, concerning the influence of the agricultural conditions

II. MATERIAL AND METHODS

The experiments were performed in the Winnogóra field station and in the experimental fields of the Institute of Plant Protection in Poznań, Poland. For the realization of these experiments, 50 m² fields were selected. A water suspension of 350 g of "Owadziak pylisty 2.4" (Zaklady Chemiczne Organica-Azot, Jaworzno, Poland), which contains lindane and HCH isomers presented in Table 1, was poured on each of the fields, and applied lindane was mixed with the soil using a tiller, passing it many times to a 10-cm depth.

The investigations of disappearance of lindane in the soil were repeated two times (throughout 2 years), applying the lindane preparation once in the spring and once in the autumn. The contaminated fields were sown thereafter with carrot: in the spring 4 weeks after the lindane application and after the autumn application the sowing was left until the spring of the next year.

III. SAMPLING

For the analysis, samples of the soil from the fields were taken directly before lindane application and 1 hr after application. Then samples of the soil were taken during the vegetation period at the corresponding terms as mentioned in Tables 2 and 3. After the autumn application of lindane, the samples were taken in the spring of the next year, when the climatic conditions caused by the thawing permitted on entry to the fields. During the spring lindane application, samples were taken 4 weeks after the application at the time of the sowing. During the entire vegetation period, the samples were taken approximately every 2 weeks.

Each time the soil samples were taken with an Engler stick with a 5-cm i.d. to a 10-cm depth. The samples were taken according to the cover method,[5] in approximately 4 L of soil. In each field, the soil sample was mixed and put into a glass jar and taken to the laboratory. Thereafter, in laboratory conditions it was dried and sieved through a 0.2-mm^2 sieve to obtain a homogeneous dried sample and stored in a glass jar at a temperature of $-20°C$ until analyzed.

The samples of plants grown in contaminated fields were taken in the terms mentioned in Tables 4 and 5. They were packed in filter paper and transported to the laboratory, where they were washed under a stream of tap water and dried with a paper filter. Then they were divided according to their physiological role and homogenized. These samples were stored in glass jars at a temperature of $-20°C$ until analyzed.

IV. DETERMINATION OF HCH ISOMERS AND HCB

Solvents used for analysis were fractionally distilled and tested by gas chromatography (GC) for the presence of interfering compounds. If such compounds were detected, the solvents were purified additionally until a gas chromatogram without interfering peaks was obtained. Anhydrous sodium sulfate and Celite 545 were of analytical grade, heated overnight at 650°C; chromium trioxide as well as acetic acid and sulfuric acid were of analytical grade.

All analytical equipment was glass to prevent contamination by undesirable substances. Glassware was washed with concentrated KOH solution and concentrated sulfuric acid and rinsed first with distilled water and then with distilled acetone and petroleum ether.

The qualitative and quantitative determinations were performed with gas chromatography (Varian Model 2100) with electron-capture detection. The separation of HCH isomers were done on the glass column 360 cm × 2 mm i.d., packed with 1.5% OV-17 + 1.95% QF-1 on Gas Chrom Q 80–100 mesh, at a temperature of 185°C. Confirmation of identity of HCH isomers from the samples was made using gas chromatography/mass spectrometry, (MS) Varian Mat 44, comparing the obtained mass spectra with those of the standard substances and with the quadrupole detection, and selecting the following masses m/e 181, 183, 219. Because significant differences in the mass fragmentography of the HCH isomers was not observed, the MS identification of HCH isomers was supported with retention times of the identified HCH isomers.

The extraction procedure and cleanup were described in detail elsewhere.[26-28]

Table 2. Contents (mg/kg) of Lindane, HCH Isomers, and HCB in the Soil After Lindane Application in the Spring

Compounds	Before the application	1 hr after	Days					
			28	56	70	94	119	150
Lindane	0.0019	0.6013	0.6042	0.3697	0.3344	0.2978	0.2686	0.1738
α-HCH	0.0003	0.0017	0.0020	0.0011	0.0011	0.0016	0.0010	0.0014
β-HCH	0.0007	0.0007	0.0008	0.0005	0.0005	0.0008	0.0007	0.0005
δ-HCH	0.0002	0.0012	0.0009	0.0004	0.0006	0.0005	0.0003	0.0004
ϵ-HCH	n.d.[a]	n.d.	n.d.	n.d.	n.d.	n.d.	n.d.	n.d.
HCB	0.0004	0.0003	0.0002	0.0002	0.0002	0.0002	n.d.	0.0002

[a] Not detected.

Table 3. Contents (mg/kg) of Lindane, HCH Isomers, and HCB in the Soil After Lindane Application in the Autumn

Compounds	Before the application	1 hr after	Days					
			167	197	223	238	309	324
Lindane	0.0018	0.6693	0.6683	0.3531	0.1548	0.1442	0.0788	0.0219
α-HCH	0.0002	0.0022	0.0026	0.0014	0.0007	0.0009	0.0006	0.0003
β-HCH	n.d.[a]	n.d.	n.d.	0.0013	0.0009	0.0011	0.0010	0.0009
δ-HCH	n.d.	0.0014	0.0010	0.0005	0.0006	0.0004	0.0005	0.0004
ϵ-HCH	n.d							

**Table 4. Contents (mg/kg) of Lindane,
HCH Isomers, and HCB in Carrot
Leaves Grown in the
Contaminated Field**

Compounds	Days after sowing		
	72	97	128
Lindane	0.1483	0.0981	0.0929
α-HCH	0.0039	0.0044	0.0050
β-HCH	n.d.[a]	n.d.	n.d.
δ-HCH	n.d.	n.d.	n.d.
ε-HCH	n.d.	n.d.	n.d.
HCB	0.0004	0.0006	0.0008

[a] Not detected.

V. RESULTS AND DISCUSSION

For this investigation, carrot was grown in the contaminated fields. This plant is characterized by a great capacity to absorb pesticides from the contaminated soil and presents extreme accumulation of organic compounds. In the spring of the first year, carrots were sown 4 weeks after the lindane application, taking into account the recommended 4-week period of phytotoxicity of lindane. Then, during the entire vegetation period, samples of soil and carrot plants were taken as mentioned in the Tables 2–5. In the autumn, carrot roots were collected from the field and stored during the winter in the cellar. In the autumn in another field, lindane was applied the same as described above. The spring of the next year, the stored carrot roots were planted in the contaminated field. The aim of this study was to obtain carrot seeds.

In the spring lindane applied to the soil during the first 4 weeks did not present diminishment of concentration. During this time because of the phytotoxicity of lindane, the plant sowing was not recommended. Four weeks after the plant sowing, the concentration of lindane in the soil had diminished to about 55% of the initial value, caused by intensive rainfall. At the end of the study period after 150 days of lindane application, the value of concentration reached about 29% of the initial value. The dynamics of disappearance of lindane depended on climatic conditions, especially on rainfall and kind of plants growing in the contaminated fields, that present a special biological dynamics of their growth. In the fields with intensive plant growth, more rapid disappearance of lindane and formed HCH isomers were observed. The phenomenon of vertical transport of lindane and HCH isomers with the water was also observed, determining the presence of their residues in the water of agricultural drainage. The above two factors were determined as primarily influenced by the loss of lindane and HCH isomers in the soil.

Lindane applied in the autumn of the second year did not show a disappearance during the winter, caused by freezing, that stopped water transport processes in the soil. After abundant snowfall and unfreezing, the vertical transport of the water caused a reduction of lindane concentration at the 223rd day up to 23%

**Table 5. Contents (mg/kg) of Lindane,
HCH Isomers, and HCB in Carrot
Roots Grown in the
Contaminated Field**

Compounds	Days after sowing		
	72	**97**	**128**
Lindane	0.2482	0.1500	0.0986
α-HCH	0.0017	0.0014	0.0011
β-HCH	n.d.[a]	n.d.	n.d.
δ-HCH	n.d.	n.d.	n.d.
ϵ-HCH	n.d.	n.d.	n.d.
HCB	n.d.	n.d.	n.d.

[a] Not detected.

of the initial concentration. The subsequent disappearance of lindane and HCH isomers in the soil depended on the intensity of plant growth and rainfall during the vegetation period. Results of chemical analysis of the soil from the investigated fields presents it as the sand soil, in which adsorption processes to the active centers of the soil dominate. The process of conjugation and binding in this kind of soil forms a minor phenomenon. Thus, the intensity of loss of lindane and HCH isomers during the study period, depended principally on the vertical transport with the water and on the intensity of adsorption through the growing plants. The analytical method used in the study allows determination of lindane, HCH isomers, and HCB as bound and free. Cleaning up the sample extracts in strong acid oxidative conditions released the pesticides from complexes with biological active substances of the soil, presenting truer results of the contamination.

The obtained results of carrot leaf contamination presented a high value of lindane and α-HCH. As time passed, these values caused by plant growth and decrease of the soil contamination diminished. During the sampling period, increase of α-HCH content can be observed that only could have its origin from bioisomerization of lindane. Moreover, the carrot leaves present content of HCB residues, increased with time, that neither can be observed in the soil nor in the carrot leaves with α-HCH and HCB, compared with the lindane content, did not exceed 6.5% of the lindane value.

In the carrot roots, the contamination diminished about 2.5 times from the first day of the sampling up to the 128th day after the sowing, caused by extensive growth of the roots and declining content of investigated compounds in the soil. Among the HCH isomers, only α-HCH was observed in the quantity almost double that in the carrot leaves. The total values of lindane and HCH isomers detected in the carrot roots were below the accepted tolerances of 0.1 mg/kg.

Investigating the gradient of contamination in the carrot roots, taking into account the part of roots serviceable in the consumption such as the peeled roots and discarded peeling of roots, superior contamination in the peelings of roots was observed. Lindane and HCH isomers were specifically accumulated in the peelings, caused by the greater lipophilicity of this part of the root. The roots

Table 6. Contents (mg/kg) of Lindane, HCH Isomers, and
HCB in the Elements of Roots Grown in the
Contaminated Field (128th day)

Compounds	Whole roots	Root peelings	Peeled roots
Lindane	0.0986	0.6290	0.0278
α-HCH	0.0011	0.0038	0.0002
β-HCH	n.d.[a]	0.0012	n.d.
δ-HCH	n.d.	0.0007	n.d.
ε-HCH	n.d.	n.d.	n.d.
HCB	n.d.	0.0001	n.d.

[a] Not detected.

Table 7. Contents (mg/kg) of Lindane, HCH
Isomers, and HCB in Carrot Seeds
Grown in the Contaminated Field

Compounds	Seeds sown	Seeds obtained
Lindane	0.0340	0.0890
α-HCH	0.0060	0.0160
β-HCH	0.0020	0.0250
δ-HCH	0.0040	0.0120
ε-HCH	n.d.[a]	n.d.
HCB	0.0010	n.d.

[a] Not detected.

when whole present slighter contamination due to biological dilution of the sample. The contamination of peelings of roots were 22.7 times greater than the peeled roots. The peeled roots contained only residual levels of lindane and α-HCH at the level accepted for consumption, despite the facts that the carrot roots had great capacity to absorb pesticides from the soil and that these carrots were grown in the fields contaminated with lindane (Table 6).

The carrot seeds from plants grown in the contaminated fields presented a specific accumulation of lindane and α-, β-, and δ-HCH (the results of which are presented in Table 7). The total content of HCH isomers in the seed reached about 60% of the lindane level. The carrot seeds do not find use in the consumption; thus from the epidemiological point of view, they do not cause any danger to consumers.

From this investigation, it can be concluded that the use of lindane according to good agricultural practice, does not endanger the consumer from elevated contamination with lindane and HCH isomer residues. Observing the values of HCH isomers in the soil and grown plants, it can be concluded that the bioisomerization processes of lindane to α-, β-, and δ-HCH, but not to ε-HCH took place according to the previous observations of Steinwandter.[17-21] However, the magnitude of this phenomenon is very restrained, and the residual levels of α-, β-, and δ-HCH did not present significant importance, still remaining at the sub-parts per billion levels.

The burden of responsibility of lindane for the high contamination of the environment and food with α-, β-, and δ-HCH and HCB did not find confirmation in the performed field studies. However, it should be underlined that the process of bioisomerization of lindane in the agricultural environment takes place, but it is necessary to consider a restrained extention of this phenomenon. The total content of HCH isomers formed from lindane in the soil fluctuates to about 2% and in the carrot leaves reached 6% of the lindane residues. Taking into account its great persistence, it did not influence significantly the contamination of agricultural products and the environment. For this contamination technical HCH is responsible; up to now this insecticide has been recommended and extensively used in many countries. This was confirmed during the monitoring study of cereals that presented the contamination with HCH isomers in samples from countries still using technical HCH in grain production and sanitation.

REFERENCES

1. Anonymous. *Manual de Agroquímicos, Químico — Farmacéuticos, Alimenticios y Biológicos Veterinarios, Vol. 1* (SARH, México: Plaguicidas, 1988).
2. Battu, R.S., P.P. Singh, B.S. Joia, and R.L. Kalra. "Contamination of Bovine (Buffalo, *Bubalus bubalis* (L.)) Milk from Indoor Use of DDT and HCH in Malaria Control Programmes," *Sci. Total Environ.* 86:281–287 (1989).
3. Benezet, H.J. and F. Matsumura. "Isomerization of Gamma-BHC to Alfa-BHC in the Environment," *Nature* 243:480–481 (1973).
4. Chessells, M.J., D.W. Hawker, D.W. Connell, and I.A. Papajcsik. "Factors Influencing the Distribution of Lindane and Isomers is Soil of an Agricultural Environment," *Chemosphere* 17(9):1741–1749 (1988).
5. Cochran, W.G. *Sampling Techniques* (New York: John Wiley & Sons, Inc., 1963).
6. Copeland, M.F. and R.W. Chadwick. "Bioisomerization of Lindane in Rats," *J. Environ. Pathol. Toxicol.* 2:737–749 (1979).
7. Engst, R., R.M. Macholz, and M. Kujawa. "Zur Kontamination der Umwelt mit Hexachlorbenzol," *Die Nahrung* 21:79–86 (1977).
8. Engst, R., R.M. Macholz, and M. Kujawa. "Recent State of Lindane Metabolism. Part II," *Res. Rev.* 72:71–95 (1979).
9. Haque, A. "Release of Bound ^{14}C-Lindane Residues from Potato Plants," *J. Pest. Sci.* 13:455–459 (1988).
10. Jagnow, G., K. Haider, and P.C. Ellwardt. "Anaerobic Dechlorination and Degradation of Hexachlorocyclohexane Isomers by Anaerobic and Facultative Anerobic Bacteria," *Arch. Microbiol.* 115:285–289 (1977).
11. Kathpal, T.S., G.S. Yadav, K.S. Kushwaha, and G. Singh. "Persistence Behavior of HCH in Rice Soil and Its Uptake by Rice Plants," *Ecotoxicol. Environ. Saf.* 15:336–338 (1988).
12. Kohnen, R., K. Haider, and G. Jagnow. "Investigation on the Microbial Degradation of Lindane in Submerged and Aerated Moist Soil," *Environ. Qual. Saf.* Suppl. 3:222–225 (1975).
13. Macholz, R. "Hexachlorocyclohexan — Isomere in der Umwelt. Ursachen für die selective Kumulation von Isomeren," *Nahrung* 26:747–757 (1982).
14. Newland, L.W., G. Chester, and G.B. Lee. "Degradation of Gamma-BHC in Simulated Lake Impoundments as Affected by Aeration," *J. Water Pollut. Cent. Fed.* 41:174–188 (1969).

15. Singh, P.P., R. Singh Battu, and R. Lal Kalra. "Insecticide Residues in Wheat Grains and Straw Arising from Their Storage in Premises Treated with BHC and DDT under Malaria Control Program," *Bull. Environ. Contam. Toxicol.* 40:696–702 (1988).

16. Singh, G., T.S. Kathpal, W.F. Spencer, G.S. Yadav, and K.S. Kushwaha. "Dissipation Behavior of Hexachlorocyclohexane Isomers in Flooded Rice Soil," *J. Environ. Sci. Health* B24(4):335–348 (1989).

17. Steinwandter, H. "Zum Lindanmetabolismus an Pflanzen. I. Bildung von Hexachlorbenzol," *Chemosphere* 2:119–125 (1976).

18. Steinwandter, H. "Zum Lindanmetabolismus an Pflanzen. II. Bildung von alfa-HCH," *Chemosphere* 4:221–225 (1976).

19. Steinwandter, H. "Beitrage zum Lindanmetabolismus in der Ökosphare," *Sonderd. Landwirtsch. Forsch.* 33(2):208–215 (1976).

20. Steinwandter, H. "Beiträge zur Umwandlung der HCH-Isomere durch Einwirkung von UV — Strahlen. I. Isomerisierung des Lindans in alfa-HCH," *Chemosphere* 4:245–248 (1976).

21. Steinwandter, H. "Experiments on Lindane Metabolism in Plants. III. Formation of beta-HCH," *Bull. Environ. Contam. Toxicol.* 20:535–536 (1978).

22. Steinwandter, H. and H. Schluter. "Experiments on Lindane Metabolism in Plants. IV. A Kinetic Investigation," *Bull. Environ. Contam. Toxicol.* 20:174–179 (1978).

23. Ulman, E. *Lindane. Monographie eines insektiziden Wirkstoffs* (Freiburg, Germany: Verlag K. Schillinger. 1973).

24. Ulman, E. *Lindane. II. Supplement.* (Freiburg, Germany: Verlag K. Schillinger, 1976).

25. Vonk, J.W. and J.K. Quirijns. "Anaerobic Formation of Alfa-Hexachlorocyclohexane from Gamma-Hexachlorocyclohexane in Soil and by *Escherichia coli*," *Pest. Biochem. Physiol.* 12:68–74 (1979).

26. Waliszewski, S. and Rzepczyński, M. "Bestimmung von Rückständen von Organochlorinsecticiden im Boden. I. Bestimmung von Alfa-, Beta-, Gamma-, Delta-, Epsilon-HCH und HCB," *Fresenius Z. Anal. Chem.* 301:32 (1980).

27. Waliszewski, S.M. and G.A. Szymczyński. "Determination of Organochloroinsecticide Residues. III. GC-Determination of Alfa-, Beta-, Gamma-, Delta-, and Epsilon-BHC and HCB in Root Vegetables (Potato Tubers, Sugar Beet Roots, Carrot Roots)," *Fresenius Z. Anal. Chem.* 311:127–128 (1982).

28. Waliszewski, S.M. and G.A. Szymczyński. "Inexpensive, Precise Method for the Determination of Chlorinated Pesticide Residues in Soil," *J. Chromatogr.* 321:480–483 (1985).

29. Wegorek, W. *Zalecenia orchrony roślin na rok 1983 dotyczace chemicznego zwalczania chrób, szkodników oraz chwastów roślin uprawnych* (Poznań, Poland: Instytut Ochrony Roślin, Zaklad Upowszechniania Postepu, 1983).

CHAPTER 15

Ecological Test Procedures for Organic Xenobiotics in Terrestrial Systems

I. Scheunert, U. Dörfler, P. Schneider, R. Schroll, and A. Zsolnay

I. INTRODUCTION

In order to assess the behavior and fate of pesticides in terrestrial ecosystems, in addition to field experiments under natural conditions, laboratory test procedures under controlled conditions are indispensable. Individual processes involved in the dynamics of residue behavior in soil have to be identified and simulated in the laboratory, in order to quantify the role they play within the complex residue disappearance process under field conditions.

Experimental conditions must be reproducible and standardizable; results shall contribute to the interpretation of field observations and predict field behavior in dependence upon various conditions. Mobility in soil and leaching from soil, volatility from soil and plants, uptake into plants, metabolization and formation of bound residues, and mineralization are main processes governing the fate of pesticides in the soil-plant system.

In the following presentation, experimental methods to determine these processes, as developed in our institute preferably for the use of ^{14}C-labeled pesticides, will be shown together with selected examples of results obtained. The use of ^{14}C labeling implies that conversion products, as well as soil- or plant-bound residues, are included in the investigations.

0-87371-616-7/93/$0.00 + $.50

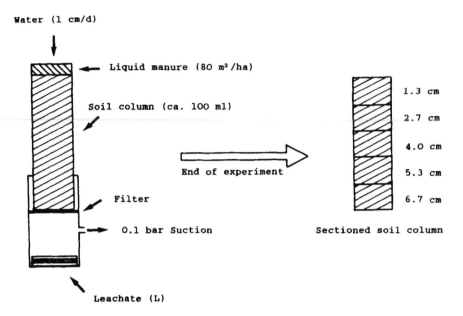

Figure 1. Laboratory soil column to study mobility and leaching of pesticides in an undisturbed soil core.

II. MOBILITY IN SOIL AND LEACHING

Figure 1 presents a laboratory soil column to study mobility and leaching of pesticides in an undisturbed soil core.[1] The pesticide is applied to the top of the soil, water is added, a slight vacuum is applied at the bottom of the column, and leachate is collected and analyzed for the pesticide and potential conversion products. At the end of the experiment, the soil core is sectioned; and each section is analyzed separately. This experimental setup is not aimed at directly predicting mobility and leaching in the field, but at obtaining relative values (e.g., at comparing different pesticides or different environmental situations).

A case study conducted with these columns was the investigation of the simultaneous presence of a herbicide and liquid manure on an acre. Figure 2 shows results obtained after the analysis of soil segments and of leachate.

The dark columns are those from experiments with [14C]atrazine and liquid manure; the light ones are from experiments with [14C]atrazine without liquid manure. It can be seen that in a peat soil having a high organic carbon content (about 12%) — as shown in the left part of Figure 2 — differences are not significant. However, in the sandy soil having a low organic carbon content (about 1%) — as shown in the right part of Figure 2 — no radioactivity is left in the upper sections of the column in the experiments without liquid manure; and in the leachate there are higher amounts of 14C without liquid manure than in the experiments with liquid manure. That means that liquid manure decreases the mobility of atrazine by increasing soil organic carbon — a fact which becomes

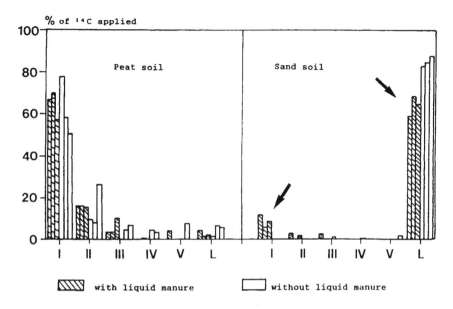

Figure 2. Mobility and leaching of [¹⁴C]atrazine in a peat and a sand soil. Soil depth: I, 1.3 cm; II, 2.7 cm; III, 4.0 cm; IV, 5.3 cm; V, 6.7 cm; L = leachate.

relevant if the original soil has a very low content of organic carbon. Thus, liquid manure may be a factor to consider in the problem of groundwater contamination by pesticides.[1]

III. VOLATILIZATION

Volatilization of pesticides from soil and plant surfaces is important for two reasons. First, volatilization contributes considerably to residue decline in soil and, therefore, is often misinterpreted as degradation. However, whereas degradation (and especially mineralization) is a mechanism for real elimination of the pesticide residue problem, volatilization only transfers the problem into another medium. Second, pesticides volatilized into the air may be transported on an over-regional scale and thus contribute to ubiquitous environmental contamination.

An apparatus to determine volatilization rates of pesticides from soil and plant surfaces must be able to keep constant all sensitive parameters influencing the volatilization process. These are primarily surface temperature, soil and air humidity, and wind velocity. In order to obtain a constant and reproducible wind velocity at every point over the surfaces, a laminar airstream is required. Furthermore, since volatilization rates of most pesticides are very low, determination must be by direct measurement according to the well-known principle that small

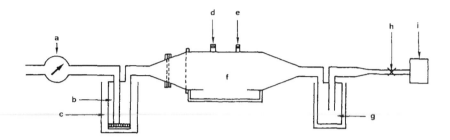

Figure 3. Apparatus to determine volatilization rates of pesticides from soil or plant surfaces. a, gas flowmeter; b, gas washing tube; c, water bath; d, thermometer; e, hygrometer; f, volatilization chamber; g, cold trap; h, valve; i, membrane pump.

Table 1. Half-Life Times of Volatilization of Lindane from Plant Surfaces

Plant	Leaf temperature (°C)	Half-life time (days)
Bean (*Phaseolus aureus*)	20	0.56
Turnip	20	0.40
Oats	20	0.31
Turnip	11	0.68
Oats	11	0.59

Note: Relative air humidity: >98%; wind velocity: 0.05 m/sec).

quantities should be measured directly rather than as a difference between two large quantities.

The apparatus shown in Figure 3 is designed to directly measure volatilized pesticides.[2] Pesticides are applied to soil or plant surfaces in a volatilization chamber, and air is drawn over the surfaces by a pump. A nearly laminar flow is achieved by the form of the chamber, the form of the air inlet funnel, and three tissue sieves within the air inlet funnel. The volatilized amounts are collected in cold traps standing either in acetone/dry ice or in a cryostat. Temperature is standardized, and drying of the soil or the plants is prevented by premoistened air. If the volatilization follows first-order kinetics, half-life times of volatilization may be calculated.

Table 1 shows, as an example, the half-life times of lindane volatilizing from leaves. Half-life times increase with decreasing temperatures and differ for different plant species due to different leaf surface anatomy.[2] The short volatilization half-life times of lindane may be an explanation for the ubiquitous presence of lindane in the air and in remote samples far away from lindane application sites.

IV. UPTAKE FROM SOIL INTO PLANTS

Uptake of pesticides from soil into plants is another process important in relation to environmental quality. Although this process does not contribute

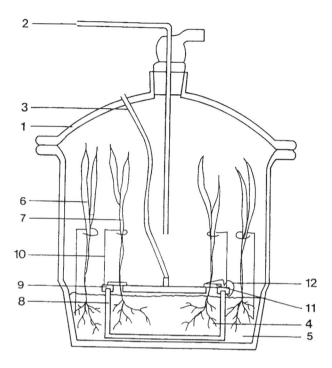

Figure 4. Closed apparatus for testing pesticides in soil/plant systems. 1, desiccator; 2, desiccator ventilation; 3, Teflon tube; 4, control soil; 5, test soil; 6, test plant; 7, control plant; 8, control dish; 9, Teflon plate; 10, steel wire; 11, small Teflon pieces; 12, metal cramps.

significantly to the disappearance of pesticide residues from soil, it contributes to pesticide residues in plants and thus to the quality of human food. Pesticides are taken into plants both via roots from the soil and via leaves from the air where they have been volatilized from the soil. Resulting total residues in the plants may be predicted only if both pathways and their dependence on pesticide properties and on environmental conditions are investigated separately.

Figure 4 shows a closed aerated laboratory apparatus where root and foliar uptake of pesticides by plants may be determined separately.[3] In addition to plants grown in soil treated with ^{14}C-labeled pesticides (here called test or t-plants), further plants are grown in untreated soil which is protected by a Teflon plate from being contaminated by pesticide vapors in the air. By these plants (here called control or c-plants), the uptake of pesticide vapor from the air by leaves may be quantified. The atmosphere in the apparatus must be analyzed carefully both for the pesticide and for $^{14}CO_2$, in order to avoid mis-interpretation of $^{14}CO_2$ uptake for pesticide uptake. In addition, ^{14}C in the plants must be characterized.

Table 2 gives results obtained for hexachlorobenzene, a model for a highly lipophilic compound.[3] The roots of the plants grown in untreated soil (c-roots)

Table 2. Concentration of [14C]Hexachlorobenzene Residues in Different Parts of Plants

	t-Roots		t-Shoots		c-Shoots	
	fm (ppm)	dm (ppm)	fm (ppm)	dm (ppm)	fm (ppm)	dm (ppm)
Oats	0.9	7.2	2.2	26.7	3.7	29.8
Maize	0.8	10.5	1.4	20.8	1.7	25.1
Lettuce	13.7	246.0	1.1	27.7	1.1	27.6
Carrots	31.6	450.0	20.6	168.5	24.3	202.0
Rape	3.2	30.3	1.4	26.4	1.4	26.7

Note: t-Roots = roots of plants grown in treated soil; t-shoots = shoots of plants grown in treated soil; c-shoots = shoots of plants grown in nontreated soil; fm = fresh plant matter; dm = dry plant matter.

contained no radioactivity. The shoots of c-plants, as shown in the right column, contained a comparable level of radioactivity as the treated t-shoots. This means that there is no translocation of hexachlorobenzene from leaves to roots and vice-versa within the time period of the experiment. In addition to this result, Table 2 shows marked differences between the plant species tested both in roots and in shoots. Carrots exhibit an outstanding behavior; their uptake of a lipophilic chemical such as hexachlorobenzene exceeds that of other plants by a factor of 10 or more.

A mathematical model for the uptake of pesticides by plants, based on this experimental model, has been published recently.[4]

V. CONVERSION AND BOUND RESIDUES

Conversion reactions of pesticides in soil or plants are of outstanding importance since by these reactions new xenobiotics may be formed, having different physicochemical properties and thus different mobility in terrestrial systems than the parent compounds, and often having unknown ecotoxicological properties. Numerous techniques to isolate these compounds and to elucidate their chemical structure will not be discussed in the framework of this presentation. However, a novel technique to isolate and identify so-called bound pesticide residues in soil and plants will be briefly presented.

Bound pesticide residues normally are called those residues which are not extractable by techniques commonly used in residue analysis. The technique of so-called supercritical fluid extraction solubilizes considerable portions of strongly bound pesticide residues. Supercritical fluids are those which at a temperature higher than the boiling point remain in the liquid state due to higher pressure. In our apparatus, an elevated temperature (250°C) is generated by a gas chromatographic oven, and an elevated pressure (150 atm) by high-performance liquid chromatography (HPLC) equipment.[5]

It is important to carefully investigate every potential formation of artifacts under these conditions and to interpret the results accordingly. Figure 5 presents,

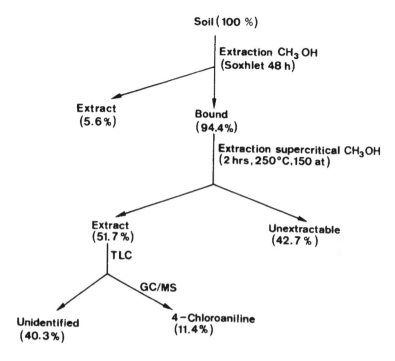

Figure 5. Extraction of soil-bound residues of monolinuron by supercritical methanol (after 16 years).

as an example, the solubilization of soil-bound aged residues of the phenylurea herbicide monolinuron by supercritical methanol.[5] After 16 years in a lysimeter, about 94% of the residues of the herbicide monolinuron in soil are bound and unextractable even by exhaustive methods such as 48-hr Soxhlet extraction with methanol. By supercritical methanol, more than half of this bound fraction was solubilized and partly identified as the metabolite 4-chloroaniline (Figure 5).

This finding gives an explanation for the fact that 4-chloroaniline was detected as a main metabolite of phenylurea herbicides in pure microbial cultures, but rarely in soils. It may be concluded that after its formation 4-chloroaniline is tightly bound to soil constituents. The portion that is not solubilized even by supercritical fluids probably is bound to soil humic matter by covalent bonds. Model experiments to demonstrate the formation of such bonds have been conducted in our institute, in cooperation with Adrian et al.[6]

VI. BIOMINERALIZATION

It is evident that determination of CO_2 resulting from mineralization of pesticides is possible only with ^{14}C-labeled pesticides, since otherwise CO_2 originating from the pesticide cannot be distinguished from that of normal soil res-

Table 3. Biodegradation of ^{14}C-labeled Pesticides 7 Days after Application to Soil

^{14}C-Labeled chemical	^{14}CO$_2$
2,4,6-Trichlorophenol	20.8
Pentachlorophenol	4.1
Aldrin	1.6
4-Chloroaniline	1.1
Atrazine	0.5
Kelevan	0.1
Kepone	0.1
Hexachlorobenzene	0.1
Dieldrin	0.1
p,p'-DDT	<0.1

Note: In % of ^{14}C applied.

piration. In the apparatus shown in Figure 4, ^{14}CO$_2$ may be collected in traps containing scintillation liquid and an organic base. The CO$_2$ traps are preceded by traps collecting organic volatiles such as the unchanged parent compounds or their volatile organic metabolites.

Table 3 presents some examples of persistent pesticides.[7-9] It can be seen that even for pesticides called highly persistent, small amounts of ^{14}CO$_2$ can be determined within 7 days. This demonstrates the high sensitivity of the experimental setup presented.

VII. CONCLUSION

In summary, it may be concluded that experimental methods presented in this chapter demonstrate actual possibilities, but also limitations in the quantification of processes determining the fate of pesticides in the terrestrial environment. Results aim at contributing to the prediction of residual behavior of pesticides in terrestrial systems in the future.

ACKNOWLEDGMENT

We thank the German Federal Environmental Agency (UBA) for giving financial support for part of this work within the framework of project 106 09 008/01 "Fate of Pesticides in the Environment — Exposure, Bioaccumulation, Degradation — Part A."

REFERENCES

1. Zsolnay, A. "Einfluss von Rindergülle auf den Transport von Atrazin in Böden." *Mitt. Dtsch. Bodenkd. Ges.* 59:511–514 (1989).
2. Scheunert, I. and U. Dörfler. Verbleib von Pflanzenschutzmitteln in der Umwelt — Exposition, Bioakkumulation, Abbau — Teil A. Forschungsbericht 106 09 008/01, Umweltbundesamt, Berlin (1989).
3. Schroll, R., I. Scheunert, and F. Korte. Determination of mass balance and of the uptake of ^{14}C-labelled pesticides from soil by plants," in *Methodological Aspects of the Study of Pesticide Behaviour in Soil,* P. Jamet, Ed. (Paris: INRA, 1989), pp. 79–85.
4. Trapp, S., M. Matthies, I. Scheunert, and E.M. Topp. "Modeling the Bioconcentration of Organic Chemicals in Plants," *Environ. Sci. Technol.* 24:1246–1252 (1990).
5. Scheunert, I., R. Schroll, and U. Dörfler. "Extraction of soil-bound residues of ^{14}C-labeled phenylurea herbicides by supercritical methanol and their identification," in Seventh International Congress of Pesticide Chemistry, Book of Abstracts, Vol. 3, H. Frehse, E. Kesseler-Schmitz, and S. Conway, Eds. (Hamburg: IUPAC/GDCh, 1990), p. 257.
6. Adrian, P., E.S. Lahaniatis, F. Andreux, M. Mansour, I. Scheunert, and F. Korte. "Reaction of the Soil Pollutant 4-Chloroaniline with the Humic Acid Monomer Catechol," *Chemosphere* 18:1599–1609 (1989).
7. Scheunert, I., B. Chen, and F. Korte. "Fate of 2,4,6-trichlorophenol-^{14}C in a laboratory soil-plant system," *Chemosphere,* 19:1715–1720 (1989).
8. Kloskowski, R. "Konzeption und Optimierung eines Pflanze/Boden Testsystems zur Bewertung von Umweltchemikalien," Doctoral Thesis, Technical University, Munich, Germany (1981).
9. Topp, E.M. "Aufnahme von Umweltchemikalien in die Pflanze in Abhängigkeit von physikalisch-chemischen Stoffeigenschaften," Doctoral Thesis, Technical University, Munich, Germany (1986).

CHAPTER 16

Role of Microbial Competition on Activity of 2,4-D Degrading *Alcaligenes xylosoxidans* Strain Introduced into Fumigated Soil

Gunalan, M.-P. Charnay, and J.-C. Fournier

I. INTRODUCTION

Contamination of the environment with residual pesticides causes serious problems. These problems may be solved by inoculation with microbial strains that are capable of degrading recalcitrant toxicants. In certain laboratories novel strains have been constructed to accelerate the biodegradation of parathion,[1,2] 2,4,5-trichlorophenoxyacetic acid,[3,4] and kelthane in soil.[5] However, the pesticide degrading capacity of such strains might decline after their release into natural ecosystems.

The *Alcaligenes xylosoxidans* strain capable of degrading 2,4-dichlorophenoxyacetic acid (2,4-D) survived well in a soil free of 2,4-D, but this strain lost its 2,4-D degrading capacity within 15 days after inoculation.[6] In fact this loss of 2,4-D degrading capacity was due to the soil biological factors that could be destroyed by various biocidal treatments before inoculation.[7]

The present study was undertaken to determine the soil biological factors which influence the loss of 2,4-D degrading capacity of the *A. xylosoxidans* strain in regard to microbial competition. These studies were aided by the use of chloroform-fumigated soil to allow the ecological studies of the inoculated strain.

0-87371-616-7/93/$0.00 + $.50

II. MATERIALS AND METHODS

A. xylosoxidans subspecies *denitrificans* which is able to use 2,4-D as the sole carbon source was grown in the 2,4-D medium containing 2.268 g of KH_2PO_4 5.920 g of Na_2PO_4, 0.123 g of $MgSO_4 \cdot 12H_2O$, 0.028 g of $FeSO_4 \cdot 7H_2O$, 0.016 g of $ZnSO_4 \cdot 7H_2O$, 0.174 g of K_2SO_4, 1.500 g of NH_4NO_3, and 1.600 g of 2,4-D/L of deionized water. The bacterium was grown for 6 days in the dark at 28°C in 100 mL of the 2,4-D medium contained in a 500 mL flask placed on a rotary shaker operating at 120 revolutions per minute. The cells were collected by centrifugation at 10,000 gravity (*g*) for 15 min at 4°C, washed and resuspended in sterile phosphate buffer saline solution (11.92 g of Na_2HPO_4, 1.92 g of NaH_2PO_4, and 8.76 g of NaCl/L of deionized water).

The soil used for this study was Dijon silty clay soil (pH 7.4, 2.1% organic matter, water-holding capacity 23.4%, 300-mg C-biomass per kilogram of soil). It was sampled to a depth of 15 cm, and passed through a 5-mm sieve before use. Fresh soil samples of 50 g (dry weight [DW]) each were placed in 250 mL bottles and exposed at 20°C for 24 hr to alcohol-free chloroform vapor in a desiccator; then the fumigant was removed from the soil samples by 10 evacuations.

To determine the effect of microbial competition on the activity of *A. xylosoxidans* in fumigated soil, two types of microbial soil suspensions — a mixture of 15 species of unidentified soil fungi and a mixture of 20 species of unidentified soil bacteria — were used as sources of microbial competitors for soil treatments. Two volumes of sterile deionized water were added to 50 g (DW) of natural soil which has been incubated for 6 days in the dark at 28°C, and the mixture was shaken for 6 hr. Soil was then allowed to settle overnight. One portion of the liquid which had been removed directly from above the settle soil was designated as a suspension of mixed soil microorganisms. Another portion of soil suspension which had been centrifuged at 6000 *g* for 10 min and passed through a sterile 3.0-μm filter was used as a suspension of soil bacteria.

The fumigated soil samples were divided into three series for three separate experiments. The first series was treated with sources of microbial competitors and inoculated with *A. xylosoxidans* immediately after fumigation. The second series was incubated for 15 days in the dark at 28°C, treated with sources of microbial competitors, and inoculated with *A. xylosoxidans*. The third series was treated with sources of microbial competitors immediately after fumigation, incubated for 15 days in the dark at 28°C, and inoculated with *A. xylosoxidans*.

Activity of the inoculated strain was expressed as the 2,4-D radioactive (specific activity 925 MBq/mmol, labeled in carbon of carboxylic group) degrading capacity. To study this activity, 2,4-D (3 μg/g of soil) was incorporated into soil samples at 0, 15, and 30 days after inoculation. The soil moisture was adjusted to 100% of water-holding capacity. The soil samples were placed in a 2 L hermetic closing jar and incubated in the dark at 20°C. The degradation of

Table 1. Evolution of 2,4-D Degrading Capacity of *A. xylosoxidans*[a] in Fumigated Soil Which Was Treated with Sources of Microbial Competitors and Inoculated with *A. xylosoxidans* Immediately after Fumigation

	% of degradation at 24 hr after 2,4-D incorporation[b]		
Soil treatments	2,4-D 0 day after inoculation	2,4-D 15 days after inoculation	2,4-D 30 days after inoculation
Untreated	47.13 ± 0.62	58.17 ± 0.37	45.31 ± 0.90
Mixture of 15 species of unidentified soil fungi	47.06 ± 1.54	51.23 ± 1.97	40.77 ± 0.57
Mixture of 20 species of unidentified soil bacteria	43.73 ± 1.12	45.53 ± 0.62 (51.64 ± 0.85)[c]	19.06 ± 1.12 (37.61 ± 0.80)[d]
Suspension of mixed soil microorganisms	43.00 ± 2.04	46.00 ± 0.62 (54.21 ± 0.75)[c]	23.32 ± 0.85 (39.65 ± 1.52)[b]
Suspension of soil bacteria	43.10 ± 2.04	47.05 ± 1.57 (57.13 ± 1.27)[c]	30.11 ± 0.67 (44.44 ± 1.39)[b]

[a] Initial concentration of inoculant = 2.5×10^7 bacteria per gram of dry soil.
[b] Mean ±95% confidence interval.
[c] Addition of glucose (1 mg/g of soil) at 13 days after inoculation.
[d] Addition of glucose (1 mg/g of soil) at 28 days after inoculation.

2,4-D radioactive in soil was observed at 24 hr after its incorporation. The evolved $^{14}CO_2$ was trapped in 10 mL of 0.2 N sodium hydroxide, mixed with 10 mL of scintillation cocktail, and measured with a Packard Tricarb 460C liquid scintillation counter.

III. RESULTS

In the first experiment, the fumigated soil was treated with the four different sources of microbial competitors and inoculated with *A. xylosoxidans* immediately after fumigation. In these conditions (Table 1), 2,4-D degrading capacity of the inoculated strain at 15 days after inoculation in all samples was slightly higher than its initial capacity.

The 2,4-D degrading capacity of *A. xylosoxidans* at 30 days after inoculation was maintained with little decline in the samples treated with a mixture of soil fungi. In contrast it markedly decreased in samples treated with the other microbial sources (a mixture of soil bacteria or a suspension of mixed soil microorganisms or a suspension of soil bacteria). For example, the 2,4-D degradation at 24 hr after incorporation in the untreated samples and in the samples treated with a mixture of soil bacteria was 40.77 and 19.06%, respectively. The addition of glucose (1 mg/g of soil) into such treated samples at 28 days after inoculation resulted in an increase of 2,4-D degrading at 30 days after inoculation up to 37.61%.

In the second experiment, the fumigated soil was incubated before treating with sources of microbial competitors and inoculated with *A. xylosoxidans*. In these conditions (Table 2), the 2,4-D degrading of the inoculated strain at 15

Table 2. Evolution of 2,4-D Degrading Capacity of *A. xylosoxidans*[a] In Fumigated Soil Which Was Incubated for 15 Days in the Dark at 28°C, Treated with Sources of Microbial Competitors, and Inoculated with *A. xylosoxidans*

Soil treatments	% of degradation at 24 hr after 2,4-D incorporation[b]		
	2,4,-D 0 day after inoculation	2,4-D 15 days after inoculation	2,4-D 30 days after inoculation
Untreated	45.93 ± 1.32	40.13 ± 0.57	36.77 ± 0.77
Mixture of 15 species of unidentified soil fungi	44.57 ± 0.75	34.36 ± 0.92	28.87 ± 1.37
Mixture of 20 species of unidentified soil bacteria	41.73 ± 1.15	13.72 ± 0.50 (34.39 ± 1.17)[c]	8.05 ± 0.40 (17.72 ± 0.85)[d]
Suspension of mixed soil microorganisms	43.22 ± 0.55	15.03 ± 0.67 (32.02 ± 0.97)[c]	8.54 ± 0.17 (17.26 ± 0.87)[d]
Suspension of soil bacteria	44.00 ± 0.50	15.50 ± 0.40 (32.06 ± 0.95)[c]	9.38 ± 1.30 (17.39 ± 1.42)[d]

[a] Initial concentration of inoculant = 2.5×10^7 bacteria per gram of dry soil.
[b] Mean ±95% confidence interval.
[c] Addition of glucose (1 mg/g of soil) at 13 days after inoculation.
[d] Addition of glucose (1 mg/g of soil) at 28 days after inoculation.

and 30 days after inoculation decreased somewhat in the samples treated with a mixture of soil fungi. In contrast, this capacity abruptly decreased in samples treated with the other microbial sources. For example, the 2,4-D degradation in the untreated samples and in the samples treated with a suspension of mixed soil microorganisms at 15 days after inoculation was 40.13 and 15.03%, respectively. When 1 mg of glucose per gram of soil was added to these treated samples at 13 days after inoculation, the 2,4-D degradation at 15 days after inoculation could be ameliorated. In this instance the percentage of 2,4-D degradation was 34.39%.

In the third experiment, the fumigated soil was treated with sources of microbial competitors immediately after fumigation, incubated for 15 days, and then inoculated with *A. xylosoxidans*. In these conditions (Table 3), the 2,4-D degrading capacity of the inoculated strain at 15 and 30 days after inoculation decreased little in the samples treated with a mixture of soil fungi. In contrast, it was totally inhibited by the other microbial sources. For example, the 2,4-D degradation in the untreated samples and in the samples treated with a suspension of soil bacteria at 15 days after inoculation was 52.77 and 1.08%, respectively. A glucose amendment (1 mg/g of soil) at 13 days after inoculation resulted in a little amelioration of 2,4-D degradation at 15 days after inoculation in these treated samples, 9.39 vs 1.08%.

IV. DISCUSSION

Sterilization of soil permitted rapid growth of inoculants.[8,9] It was most probably attributable to the death of indigenous microorganisms capable of competing

Table 3. Evolution of 2,4-D Degrading Capacity of *A. xylosoxidans*[a] in Fumigated Soil Which Was Treated with Sources of Microbial Competitors Immediately After Fumigation, Incubated for 15 Days in the Dark at 28°C, and Inoculated with *A. xylosoxidans*

Soil treatments	% of degradation at 24 hr after 2,4-D incorporation[b]		
	2,4-D 0 day after inoculation	2,4-D 15 days after inoculation	2,4-D 30 days after inoculation
Untreated	57.01 ± 0.75	52.77 ± 1.62	47.44 ± 1.12
Mixture of 15 species of unidentified soil fungi	50.82 ± 0.77	46.45 ± 1.02	35.32 ± 1.15
Mixture of 20 species of unidentified soil bacteria	46.12 ± 0.35	1.06 ± 0.52 (9.55 ± 1.07)[c]	0.56 ± 0.17 (5.14 ± 0.15)[d]
Suspension of mixed soil microorganisms	48.54 ± 1.17	0.96 ± 0.25 (9.09 ± 2.09)[c]	0.53 ± 0.17 (4.41 ± 0.85)[d]
Suspension of soil bacteria	49.46 ± 1.77	1.08 ± 0.37 (9.39 ± 1.34)[c]	0.68 ± 0.45 (5.54 ± 1.29)[d]

[a] Initial concentration of inoculant = 2.6×10^7 bacteria per gram of dry soil.
[b] Mean ±95% confidence interval.
[c] Addition of glucose (1 mg/g of soil) at 13 days after inoculation.
[d] Addition of glucose (1 mg/g of soil) at 28 days after inoculation.

for the limited nutrient resources of soil.[10] These killed natural microflora might act as nutrient sources for the inoculants.[11] For a short period following a soil sterilization certain soil organisms, particularly bacteria, increased in populations sizes.[12] It was associated with an increase in the consumption of oxygen and the mineralization of carbon, nitrogen, and other elements.[13]

In this study, we attempted to determine the role of microbial competition on activity of an inoculated strain in varying soil conditions allowing the growth of both microorganisms. The first condition permitted the microbial competitors and the inoculated strain to proliferate, since they were immediately introduced into soil after fumigation. In the second condition, the growth of both microorganisms was more limited than in the first condition, inasmuch as they were introduced into fumigated soil which had been incubated for 15 days in the dark at 28°C before the treatments. Finally, the third condition provided a proliferation for the microbial competitors. This condition gave them an advantage over the inoculated strain, because they were immediately introduced into soil after fumigation. On the other hand, the inoculated strain was inoculated 15 days after the introduction of the microbial competitors.

The decline of 2,4-D degrading capacity of the inoculated strain apparently depended on its ability to compete with microbial competitors in relation to the available nutrients. For this reason, the incorporation of microbial sources into samples containing large availability of nutrients did not result in a considerable decrease in 2,4-D degrading capacity of *A. xylosoxidans* after inoculation. On the contrary, this capacity markedly decreased in the lacking nutrients samples which were treated with microbial sources. Among these microbial sources, we noticed that the most important effect was obtained with bacterial sources. The soil fungi did not cause an appreciable decrease in activity of the inoculated strain. The suspension of mixed soil microorganisms should have to contain both bacteria and the other microorganisms, e.g., actinomycetes, protozoa, fungi, etc. However, this soil suspension affected the 2,4-D degrading capacity of *A. xylosoxidans* in the same fashion that the bacterial sources (a mixture of soil bacteria or a suspension of soil bacteria) did. In fact, the addition of glucose in such samples at 13 and 28 days after inoculation could ameliorate the capacity of the inoculated strain to degrade 2,4-D at 15 and 30 days after inoculation. The inability of *A. xylosoxidans* to compete with indigenous microorganisms and the indirect effect of glucose to stimulate the activity of *A. xylosoxidans* after its release in soil were shown in another experiment.[14]

On the basis of these data, it is concluded that competition between soil bacteria for organic carbon is responsible for the abrupt decrease in capacity of the inoculated strain to degrade the pesticide in soil.

REFERENCES

1. Sethunathan, N. "Organic Matter and Parathion Degradation in Flooded Soil," *Soil Biol. Biochem.* 5:641–644 (1973).
2. Daughton, C.G. and D.P.H. Hsieh. "Accelerated Parathion Degradation in Soil by Inoculation with Parathion-Utilizing Bacteria," *Bull. Environ. Contam. Toxicol.* 18:48–56 (1977).
3. Chaterjee, D.K., J.J. Kilbane, and A.M. Chakrabarty, "Biodegradation of 2,4,5-Trichlorophenoxyacetic Acid in Soil by a Pure Culture of *Pseudomonas cepacia*," *Appl. Environ. Microbiol.* 44:514–516 (1982).
4. Kilbane, J.J., D.K. Chaterjee, and A.M. Chakrabarty. "Detoxification of 2,4,5-Trichlorophenoxyacetic Acid from Contaminated Soil by *Pseudomonas cepacia*," *Appl. Environ. Microbiol.* 45:1697–1700 (1983).
5. Golovleva, L.A., R.N. Pertsova, A.M. Boronin, V.M. Travin, and S.A. Kozlovsky. "Kelthane Degradation by Genetically Engineered *Pseudomonas aeruginosa* BS827 in Soil Ecosystem," *Appl. Environ. Microbiol.* 54:1587–1590 (1988).
6. Gunalan, M.-P. Charnay and J.-C. Fournier. "Starvation, Survival and Stability of 2,4-D Degrading *Alcaligenes xylosoxidans* Strain Inoculated in Soil," II Workshop Pesticides-Soils, Alicante, Spain, Abstract, (1989a).
7. Gunalan, M.-P. Charnay and J.-C. Fournier. "The Effects of Biocidal Pretreatments on the Activity of 2,4-D Degrading *Alcaligenes xylosoxidans* Strain Inoculated in Soil," II Workshop Pesticides-Soils, Alicante, Spain, Abstract, (1989b).
8. Salonius, P.O., J.B. Robinson, and E.F. Chase. "A Comparison of Autoclaved and Gamma-Irradiated Soil as Media for Microbial Colonization Experiments," *Plant Soil* 27:239–248 (1967).
9. West, A.W., H.D. Burges, and T.J. Dixon. "Survival of *Bacillus thuringiensis* and *Bacillus cereus* spore inocula in soil: effects of pH, moisture, nutrient availability and indigenous microorganisms," *Soil Biol. Biochem.* 17:657–665 (1985).
10. Jenkinson, D.S. "The Fate of Plant and Animal Residues in Soil, in *The Chemistry of Soil Processes*, D.J. Greenland and M.H.B. Hayes, Eds. (London: John Wiley & Sons, 1981), pp. 505–561.

11. Jenkinson, D.S. and D.S. Powlson. "The Effects of Biocidal Treatments on Metabolism in Soil. V. A Method for Measuring Soil Biomass," *Soil Biol. Biochem.* 8:209–213 (1976).
12. Powlson, D.S. "Effects of Biocidal Treatments on Soil Organisms, in *Soil Microbiology,* N. Walker, Ed. (London: Butterworth Publishers, 1975), pp. 193–223.
13. Jenkinson, D.S. "Studies on Decomposition of Plant Material in Soil. II. Partial Soil Sterilization and the Soil Biomass," *J. Soil Sci.* 17:280–302 (1966).
14. Gunalan, M.-P. Charnay and J.-C. Fournier. "Effect of Glucose on Growth and Activity of 2,4-D Degrading *Alcaligenes xylosoxidans* Strain in Natural Soil, Fumigated Soil and Antibiotic Amended Soil," in *Study and Prediction of Pesticides Behavior in Soils, Plants and Aquatic Systems,* M. Mansour, Ed., GSF-Forschungszentrum fur Umwelt und Gesundheit, Munchen-Neuherberg, pp. 194–200 (1990).

CHAPTER 17

Thermodynamic Properties of Halogenated Dibenzo-p-Dioxins, Dibenzofurans, and Pesticides

B. F. Rordorf, B. Nickler, and C. M. J. Lamaze

ABSTRACT

Vapor pressures are important parameters for environmental fate modeling of chemicals. Vapor pressure curves were obtained for 80 pesticides using an automatized gas saturation method with on-line gas chromatography (GC) analysis. Special consideration was given to a variety of phenomena of importance to the outcome of vapor pressure results: sample contamination by impurities or amorphous phase, solid- to plastic-phase transitions, crystal modifications, and measurements on free acids and on labile pesticides. A vapor pressure correlation method has been developed in the past for halogenated dibenzo-p-dioxins and dibenzofurans and has been used to predict vapor pressures for 29 halogenated dibenzo-p-dioxins and for 55 chlorinated dibenzofurans. The present pesticides were chosen to contain a number of common structure elements with these compounds measured in the past, and this data should be valuable for developing vapor pressure estimation methods.

I. INTRODUCTION

While vapor pressure data has been collected in agrochemical handbooks and manuals over the years, most data is not traceable to scientific publications. Full

0-87371-616-7/93/$0.00 + $.50

vapor pressure curves are required to deduce enthalpies and entropies of sublimation (evaporation). However, they are rarely available for pesticides. Assessment and comparison of such data is complicated by the variety of vapor pressure methods used. Direct comparability of the experimental data is of importance when the determined thermodynamic parameters are used in correlation methods.

Vapor pressure measurements on 36 halogenate dibenzo-p-dioxins and dibenzofurans have been reported in the past.[3] Additional curves have been obtained for 1,6-DCDD; 1,2,3-T$_3$CDD; 1,7,8-T$_3$CDD; 2,3,7-T$_3$CDD; 1,2,7,8-T$_4$CDD; 1,3,7,8-T$_4$CDD; 1,2,4,7,8-PCDD; OCDD (octachlorodibenzo-p-dioxin); 2,8-DCDF; 1,3,7-T$_3$CDF; 2,4,7-T$_3$CDF; OCDF (octachlorodibenzofuran); 2,7-dichloro-9-fluorenone; 2,4,7-trichloro-9-fluorenone; and 2,4,7-trichlorofluorene. Some of these new compounds are related by constant chlorine substitution patterns.

II. EXPERIMENT

Vapor pressure curves were obtained by an automatized gas saturation method with on-line GC analysis.[4] Enthalpies of fusion and melting points were measured by differential scanning calorimetry (DSC) using a Perkin-Elmer DSC-2 with a data station 3500.

III. RESULTS AND DISCUSSION

A few selected vapor pressure curves are displayed in Figure 1. The vapor pressure curves of octachlorodibenzo-p-dioxin and of octachlorodibenzofuran have been remeasured.[4] Somewhat different values for the molar enthalpies and entropies of sublimation have been obtained from these new recordings: h_s = 149822 J/mol and s_s = 311.8 J/mol K (120–200°C) for OCDD (Table 1, bottom), and h_s = 143702 J/mol and s_s = 308.3 J/mol K (105 to 261°C) for OCDF (Table 1, top).

Extrapolated 25°C vapor pressure values for halogenated dibenzo-p-dioxins, dibenzofurans, and pesticides are compared in Figure 2. Extrapolations involved, in some of the cases, a recalculation from the liquid- to solid-phase curves using experimental enthalpies of fusion from the DSC measurements. The dioxin and furan values show a clear-cut correlation with the molecular weight. In contrast to these compounds related by homology, no such correlation is observed for the pesticides.

Added confidence in vapor pressure measurements was gained by a comparison of the enthalpies of fusion obtained from DSC and values from the difference of slopes of vapor pressure curves. This comparison was possible where vapor

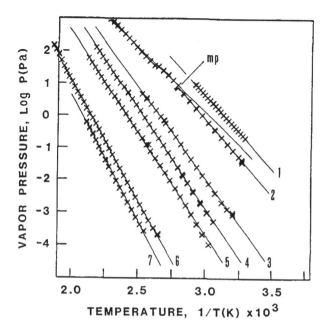

Figure 1. Vapor pressure curves of fluorene (1); 9-fluorenone (2); 2,4,7-trichlorofluorene (3); 2,4,7-trichloro-9-fluorenone (4); 1,2,4,7,8-pentachlorodibenzo-p-dioxin (5); octachlorodibenzofuran (6); and octachlorodibenzo-p-dioxin (7). The melting point (mp) of fluorenone is indicated and the linear regressions are traced for the solid and liquid phases, respectively. All of the remaining curves are for the solid compounds.

pressure measurements were accessible in both solid and liquid phases. The measurement of enthalpies of fusion (h_m) and the melting points (T_m) by DSC permitted us to test Walden's and Fishtine's rules. The calculated entropies of fusion ($s_m = h_m/T_m$) were then plotted against T_m for all of the compounds (Figure 3, bottom). A correlation close to Walden's rule (64 J/mol K compared to the usually assumed value of 56.5 J/mol K) is apparent for the rigid halogenated dibenzo-p-dioxins and dibenzofurans. The entropies of fusion for the pesticides, on the other hand, show a great amount of scatter.

Liquid-phase vapor pressure curves were not directly measured for many of the pesticides. Often the liquids were not experimentally accessible. Vapor pressure curves over the liquids were then calculated from solid-phase vapor pressure curves using the enthalpies of fusion from the DSC. These curves for the liquid compounds were then extrapolated up to the boiling point. The curvatures for these extrapolations were estimated by group contribution methods for the dioxins and furans.[1] Fixed values of delta c_p (gas minus liquid) of -70 J/mol K were assumed for the pesticides. The extrapolations were done in an iterative procedure using the integrated Clausius-Clapeyron equation by a similar method as described by Rordorf.[1]

Table 1. Recorded Vapor Pressures of Octachlorodibenzo-p-Dioxin (Bottom) and Octachlorodibenzofuran (Top)

t(°C)	mPa				t(°C)	mPa			
105.6	0.230	0.224	0.218	0.213	185.6	560	574	573	573
110.5	0.377	0.367	0.362	0.354	190.7	845	840	852	851
115.5	0.627	0.623	0.626	0.618	195.8	1094	1100	1102	1102
120.6	1.09	1.09	1.09	1.08	200.9	1633	1640	1643	1643
125.5	1.87	1.88	1.88	1.87	205.9	2401	2414	2416	2420
130.5	3.19	3.22	3.20	3.21	210.9	3513	3528	3526	3539
135.5	5.40	5.43	5.42	5.42	215.9	5105	5130	5135	5133
140.4	8.98	8.99	9.04	9.00	221.0	7353	7397	7417	7403
145.4	14.7	14.7	14.8	14.8	226.0	10627	10752	10784	10776
150.5	24.0	24.1	24.1	24.1	231.0	15136	15202	15329	15269
155.5	38.5	38.6	38.5	38.5	236.0	22294	22572	22455	22685
160.5	63.4	64.2	64.2	63.5	241.1	31556	31887	32149	32006
165.6	97.7	99.2	101	101	246.1	48268	48287	48265	48125
170.5	152	151	157	156	251.1	71468	73785	76375	75799
175.6	236	241	246	234	256.2	98377	101697	105899	105964
180.6	366	381	374	375	261.2	129504	137929	139425	139924
120.1	0.277	0.283	0.279	0.280	165.0	24.7	25.3	25.4	25.6
124.9	0.444	0.448	0.449	0.445	170.0	40.1	41.1	40.7	41.2
129.9	0.725	0.740	0.741	0.743	175.0	64.2	64.8	66.8	66.1
134.9	1.23	1.25	1.26	1.25	180.0	99.3	102	105	103
139.9	2.06	2.09	2.09	2.11	185.0	155	159	160	160
144.8	3.47	3.52	3.54	3.55	189.8	239	261	246	260

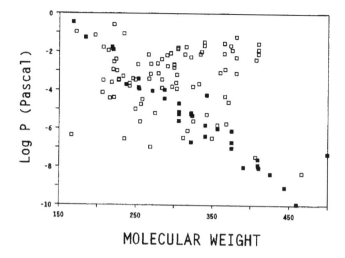

Figure 2. Extrapolated vapor pressures for 25°C of the following compounds: DD, 1-MCDD, 2-MCDD, 137-T3CDD, 2378-T4CDD, 12347-PCDD, 123478-H6CDD, 1234678-H7CDD, OCDD, 36-DCDF, 1234678-H7CDF, and OCDF where C stands for chlorine (further chlorinated DDs and DFs of Reference 3 are included in this figure but not in Figure 3). The pressures for the dibenzo-p-dioxins (DD) and dibenzofurans (DF) (filled symbols) are compared to the values for various pesticides (open symbols; see Figure 3 for a list of compounds).

These extrapolations permitted us to test Trouton's rule and in its enhanced version Fishtine's rule. The latter rule estimates entropies of evaporation at the boiling point as $s_v (Tb) = K_F (36.61 + R \ln Tb) + R \ln p°$ with the Fishtine constant $K_F = 1.01$, the gas constant $R = 8.314$ J/mol K, and $p° = 101325$ Pa. Figure 3 shows differences of the extrapolated entropies minus estimations by the Fishtine' rule as a function of molecular weight of the compounds. A great amount of scatter is apparent for both the dioxins/furans and the pesticides. A similar picture results when plotting the entropy differences against ln Tb. Some of the scatter is probably due to uncertainties of measured enthalpies of sublimation, of evaporation, and of fusion, as well as to estimated delta c_p values used.

In conclusion, we developed a large thermodynamic data set on environmental contaminants. These data permit us for the first time to test some of the widely used thermodynamic rules as applied to high molecular weight/low volatility compounds. These rules were mostly developed for much more volatile compounds. They have rarely been directly tested for organic substances of low volatility in the past. More high quality thermodynamic data are clearly needed on this class of compounds of high environmental impact.

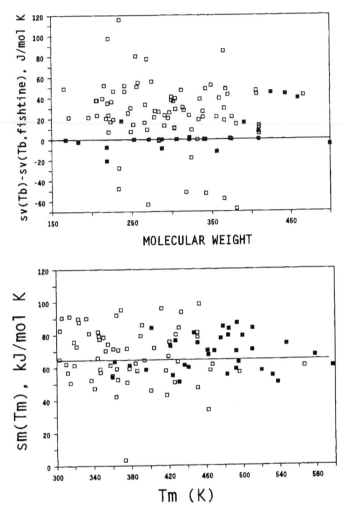

Figure 3. Bottom: entropies of fusion at the melting point (T_m, Waldens rule). Top: calculated minus predicted entropies of evaporation at the boiling point (T_b, Fishtine rule). Filled symbols stand for dibenzo-p-dioxins and dibenzofurans (Figure 2) and open symbols for these pesticides: aldrin, ametyne, atrazine, azamethiphos, azinphos-methyl, banvel, barban, bifenox, bromophos-methyl, captafol, chloranil, chlorben-side, chlorbufam, chlordane, chlordimeform, chlorfenethol, chlorfenvinphos, chlor-fenson, chloroneb, chloropham, chlorthiamid, cyromazine, chlorthal-methyl, 2,4-D, 2,4-DB, diazinon, dichlobenil, dichlofenthion, dichlorophen, dichlone, dichlor-phos, dichlorprop, dichloran, dicofol, dieldrin, dihydroheptachlor, diuron, endo-sulfan, endrin, fenchlorphos, fenoprop , fenthion, fluometuron, fthalid, genite, hep-tachlor, imazalil, isazofos, lindane, linuron, metalaxyl, methidathion, metolachlor, naled, neburon, nitrofen, penconazole, parathion, pentachlorophen, phosalone, phosphamidon, profenofos, prometon, prometryn, propanil, propazine, propicon-azole, quintozene, simazine, sulfallate, 2,4,5-T, swep, temephos, terbutryne, te-trachlorvinphos, tetradifon, tetrasul, trichloronat, triclopyr, trifluralin, and triallate.

REFERENCES

1. Rordorf, B.F., *Chemosphere* 18:783–788 (1989).
2. Rordorf, B.F., 33rd Conference of the International Association for Great Lakes Research, Windsor, Ontario, Canada (May 13 to 17, 1990).
3. Rordorf, B.F., L.P. Sarna, G.R.B. Webster, S.H. Safe, L.M. Safe, D. Lenoir, K.H. Schwind, and O. Hutzinger. 9th International Symposium on Chlorinated Dioxins and Related Compounds, Toronto (1989).
4. Rordorf, B.F., *Chemosphere* 15:1325–1332 (1986).

CHAPTER 18

New Results and Considerations on the Bioconcentration of the Superlipophilic Persistent Chemicals Octachlorodibenzo-p-Dioxin (OCDD) and Mirex in Aquatic Organisms

Harald J. Geyer and Derek C.G. Muir

I. INTRODUCTION

Bioconcentration, toxicity, and biotic and abiotic degradation of pesticides and other chemicals in aquatic organisms (such as fish) are important criteria in hazard assessment. A high bioconcentration potential of a chemical substance in biota increases the probability of toxic effects being encountered in man and his environment. Aquatic organisms may be contaminated by chemicals by several pathways: directly via uptake through gills or skin as well as indirectly via ingestion of food or contaminated sediment particles. For clarity the terminology associated with such studies should be given.[1,2]

A. Bioconcentration

Bioconcentration is the direct uptake of a chemical by an organism only from water. Experimentally, the result of such a process is reported as the bioconcentration factor (BCF). Consequently, the BCF is defined as the ratio of steady-state concentration of the chemical in aquatic organisms (C_o) and the corresponding concentration in the surrounding water (C_w):

0-87371-616-7/93/$0.00 + $.50

© 1993 by Lewis Publishers

$$BCF = \frac{C_o}{C_w} \quad \frac{[ng/kg]}{[ng/L]} \tag{1}$$

For aquatic organisms three different bioconcentration factors can be given:

1. on a wet weight basis (BCF_w)
2. on a lipid basis (BCF_L)
3. on a dry weight basis (BCF_D)

All these BCF values are dimensionless because 1 L H_2O is 1 kg. For the sake of comparison, the most important BCF value is that on a lipid basis.[26] BCF_L values can easily be calculated from BCF_w values, if the lipid content (L in % on a wet weight basis) of the organism is known:

$$BCF_L = \frac{BCF_w \cdot 100}{L\ (\%)} \tag{2}$$

BCF should be determined in an appropriate concentration range where values are independent of concentration of the compound and are ecologically meaningful and no toxic effects occur.

The definition of bioconcentration has to be distinguished from the terms of indirect contamination, such as biomagnification and bioaccumulation. Both terms are also associated with increasing concentrations of a chemical in organisms.

B. Biomagnification

Biomagnification is the direct uptake of a chemical by an organism via food. In aquatic environments, this process operates simultaneously with bioconcentration.[4]

There are different opinions as to whether and to what extent biomagnification occurs. One hypothesis declines any biomagnification. Macek et al.[3] reviewed data derived from simple experimental laboratory food chains and concluded that most chemicals are by no means biomagnified. This statement fits well the common theory of the partitioning process of lipophilic chemicals, which considers animals as simple lipid aggregations. Other authors, however, came to the conclusion that there is biomagnification of very hydrophobic chemicals.[32-33]

C. Bioaccumulation

Bioaccumulation is defined as the uptake of chemicals via food and water. Whether this hypothesis also applies for superlipophilic chemicals, is being investigated.

Table 1. Properties of Octachlorodibenzo-p-Dioxin (OCDD)

Structure

CAS-Number	3268-87-9
Mol wt (g/mol)	460
Melting point (°C)	332
Aqueous solubility (pg/L)[27,28,29]	74 (25°C)
	400 (20°C)
n-Octanol/water partition coefficient (log K_{ow})[27,38]	8.20 and 8.60
Sorption coefficient on an organic carbon basis (log K_{oc})[40]	7.90
Bioconcentration factor in aquatic organisms on a lipid basis (BCF$_L$)[a]	8.5×10^7
Bioaccumulation factor in human fat (BAF$_L$)[b]	8.0×10^4

[a] Bioconcentration factor at steady state of OCDD in fish and mussel (this study).

[b] Extrapolated bioaccumulation factor calculated from equation: $\log BAF_L = 0.745 \cdot \log K_{ow} - 1.19$. However, this steady-state value will never be reached in human fat.[41]

To start with, this chapter deals with mere bioconcentration (usually in fish) of such hydrophobic chemicals from water. The real BCF value of a chemical has to be independent of the water concentration. In all cases, however, where bioconcentration factors differ by some orders of magnitude for the same chemical, although they have been determined under nearly equal experimental conditions with fish of the same species, sex, age, body weight, and lipid content, it has to be questioned whether a "true" bioconcentration factor was found. Consequently, all other experimental conditions have to be reexamined.

We have compiled and reexamined BCF values of octachlorodibenzo-p-dioxin (OCDD) and Mirex, and try to give some explanations for the apparent dependency of the BCF values on concentrations of these superlipophilic chemicals as well as to present methods for estimating "true" bioconcentration factors.

OCDD and Mirex were selected as model chemicals for a number of reasons. Both substances are highly persistent and resistent to biotic and abiotic degradation, except for photolysis. Usually, OCDD is the most prevalent polychlorinated dibenzo-p-dioxin congener found in pentachlorophenol (PCP), fly ash, sediments, fish, and other biotic samples.[39]

OCDD belongs to the group of "superlipophilic" compounds[5] with octanol/water partitioning coefficients (log K_{ow}) between 8.2 and 8.6 and water solubilities between 74 and 400 pg/L (see Table 1). OCDD is not produced for commercial purposes and has no reported use although this chemical and other octahalogenated dibenzo-p-dioxins were proposed by a Canadian company as chemical intermediates, biocides, and flame-retardants.[34]

Mirex is an organochlorine insecticide that was used for imported fire ant control in large areas of the southeastern United States. Formulated as a bait, Mirex was intended to control the imported fire ant (*Solenopsis richteri* Forel and *Solenopsis invicta* Buren). Mirex belongs also to the group of superlipophilic persistent chemicals (log K_{OW}: 6.89).

II. BIOCONCENTRATION BIOASSAY

To measure steady-state bioconcentration factors of superlipophilic chemicals with log $K_{OW} > 6$, such as OCDD and Mirex, it is essential to determine both uptake rates (K_u) and depuration rates (K_d) because such extremely hydrophobic compounds reach steady-state concentrations only after considerably long times (several months):[36]

$$BCF = \frac{K_u}{K_d} \frac{[d^{-1}]}{[d^{-1}]} \tag{3}$$

For further information about the kinetic methods for determining bioconcentration see, for instance, References 22 and 36. Since ambient chemical concentrations must be kept relatively constant during the test, flow-through systems preferably should be applied.

III. STATE OF THE ART

For aquatic organisms, bioconcentration can be considered to be the partitioning of a chemical between the lipid phase of the organism and the ambient water at equilibrium. With dissolved chemicals having log $K_{OW} < 6$, wet weight bioconcentrations depend on K_{OW} as given in Equation 4:

$$\log BCF_w = 1.00 \log K_{OW} + \log (L/100) \tag{4}$$

where L is the lipid content in percent on wet weight basis.[6]

This linear regression should be applied to chemicals only possessing log $K_{OW} < 6$. However, for chemicals with log $K_{OW} > 6$ "parabolic" or "bilinear" relationships between log BCF and log K_{OW} have been proposed.[7,8,18]

It seems clear that steric parameters, such as molecular size and cross sections can influence the bioconcentration of hydrophobic chemicals in aquatic organisms. The hypothesis of Opperhuizen et al.[12] proposed a lack of membrane permeability, due to the large cross sections of >9.5 Å, which prevents clear bioconcentration of superlipophilic chemicals in fish. Therefore, superlipophilic substances, i.e., OCDD or Mirex®, appear to be bioconcentrated less than calculated from their n-octanol/water partitioning coefficient (K_{OW}).[5]

IV. METHOD FOR PREDICTION OF BCF$_L$ VALUE OF OCDD

Wet weight bioconcentration factors (BCF$_W$) of OCDD in various fish species were extracted from recent articles.[5,16-18] Only steady-state BCF data, which were obtained in flow through systems, were taken into consideration. For sake of comparison, BCF$_W$ values had to be transformed into BCF$_L$ values.[26] Table 2 contains body weights, lipid contents, BCF$_W$, and BCF$_L$ values, with corresponding ambient OCDD concentrations being given as well. In order to assess the most likely BCF$_L$ (ambient OCDD concentrations < water solubility), experimental BCF$_L$ data of OCDD were plotted against respective external OCDD on a log/log basis (Figure 1). The most likely BCF$_L$ value of OCDD in fish was obtained from an extrapolation of the linear relationship down to water solubility of 74 pg/L. The same procedure was applied to Mirex.

V. RESULTS AND DISCUSSION

Although not many BCF data are available for OCDD, it is obvious that the experimentally obtained BCF$_L$ in different fish species clearly depends on ambient OCDD concentrations (Table 2, Figure 1). That means that BCF$_L$ values increase with decreasing external concentration in the experiments. Using only the three highest BCF$_L$ values (center of Figure 1), which were published by Muir et al.,[16,18] a linear regression was found. Equation 5 reads:

$$\log BCF_L = 11.18 - 1.74 \cdot \log C_W \tag{5}$$

where C_W is the OCDD concentration (pg/L) in ambient water. At water solubility 74 pg/L this regression reads 8.5×10^7. This BCF$_L$ value exceeds even the published maximum value by two orders of magnitude. The same applies when considering BCF$_W$ values.

The same procedure was applied to Mirex (Figure 2). As expected, BCF$_W$ values at ambient concentrations not exceeding water solubility lie clearly above the maximum values published thus far[31]: 130,000 rather than 51,300, 12,400, or 3,700. In a flow-through system having 32 days of exposure and using a mean exposure concentration of 1,200 ng/L, Veith et al.[30] found a bioconcentration factor (BCF$_W$) of 18,200 in fathead minnows. However, all these BCF$_W$ values of Mirex are no steady-state bioconcentration factors.

Table 2. Bioconcentration Factors on a Wet Weight Basis (BCF$_w$) and on a Lipid Basis (BCF$_L$) of OCDD in Different Fish Species in Dependence on OCDD Concentrations in Ambient Water (C$_w$)

Fish species	Mean body weight (g)	Lipid content (%)	OCDD conc. (pg/L)	Bioconcentration factor		Ref.
				BCF$_w$	BCF$_L$	
Guppy (male)	0.1	3.5	4.0×10^6	<1,050	$<3 \times 10^4$	5
Rainbow trout	0.3	6.9	4.15×10^5	34	4.9×10^2	16
Rainbow trout	0.3	6.9	2.0×10^4	136	2.0×10^3	16
Guppy (female)	0.079	7.5	6.4×10^5	703	9.4×10^3	17
Fathead minnow	1.7	3.5	9.0×10^3	2,226	6.4×10^4	16
Fathead minnow	1.7	3.5	7.0×10^3	22,300	6.4×10^5	18
Fish	—	5.0[b]	7.4×10^4 [a]	4.3×10^6	8.5×10^{7c}	Present work

Source: From H. Geyer, et al. *Chemosphere* 25, 1257–1264 (1992). With permission.

[a] Water solubility of OCDD.
[b] Assumed average lipid content (% on a wet weight basis) of fish.
[c] Predicted by extrapolation from Equation 1.

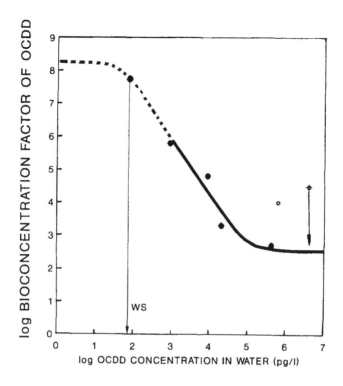

Figure 1. *Relationship between bioconcentration factor on lipid basis* (BCF_L) *of octachlorodibenzo-p-dioxin (OCDD) in fish and the OCDD concentration in ambient water (WS: water solubility of OCDD = 74 pg/L). (From H. Geyer, et al.* Chemosphere *25, 1257–1264 (1992). With permission.)*

With mussels and fish, which are contaminated via the sediment, Geyer et al.[24,25] calculated BCF values using Equation 6:

$$BCF = \frac{C_O}{C_W} = \frac{C_O \cdot K_{OC} \cdot \%OC}{C_S \cdot 100} \tag{6}$$

where %OC = organic carbon content (%) of the sediment
K_{OC} = sorption coefficient on an organic carbon content
C_S = sediment OCDD concentration on a dry weight basis

This indirect method revealed BCF values which are consistent with those obtained by the above extrapolation.

Another way in calculating BCF values of OCDD is the application of two quantitative structure activity relationships (QSARs) based on log K_{OW} (OCDD: 8.60) as well as water solubility (OCDD: 74 pg/L), which was developed for mussels (*Mytilus edulis*) on the wet weight basis by Geyer et al.:[29,35]

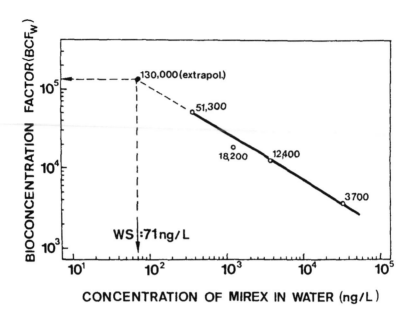

Figure 2. Relationship between bioconcentration factor on a wet weight basis (BCF$_w$) of Mirex in fish and the Mirex concentration in ambient water (WS: water solubility of Mirex = 71 ng/L). (From H. Geyer, et al. *Chemosphere* 25, 1257–1264 (1992). With permission.)

$$\log BCF_w = 0.858 \cdot \log K_{ow} - 0.808 \qquad (7)$$

$$\log BCF_w = 4.94 - 0.682 \cdot \log WS \ (\mu g/L) \qquad (8)$$

These equations give an estimate of OCDD BCF$_w$ of 3.7×10^6 and 5.7×10^7, respectively. Although both figures differ considerably, they indicate again that experimentally derived BCF values are too small by at least one or two orders of magnitude.

VI. CONCLUSIONS

First, it is evident that BCF values of superlipophilic chemicals in fish must be relatively low in cases of too short exposure times, because no steady-state can be achieved.[19]

Second, various chemicals, which do not bioconcentrate significantly, are relatively quickly metabolized or excreted. Superlipophilic chemicals mentioned earlier, however, are very resistant against metabolic transformation or microbial degradation. As a consequence, possible metabolism cannot serve as a reason for the observed low bioconcentration factors of OCDD, Mirex, and other superlipophilic persistent chemicals.

Third, in several studies OCDD has been identified to be that polychlorinated dibenzo-p-dioxin (PCDD) with the highest concentration in marine and freshwater biota, such as mussels, shrimps, and fish.[13-15,42] Muir et al.[16] and Gobas and Schrap[17] have demonstrated in bioconcentration experiments that OCDD was found in elevated concentrations in the organism, even when the gastrointestinal tract and the gills were removed. That means that such superlipophilic chemicals are able to penetrate membranes and are bioconcentrated in fish and other aquatic organisms.

Fourth, the main argument, however, against the relatively low experimentally determined bioconcentration potential of such persistent highly hydrophobic compounds except for 2,3,7,8-tetrachlorodibenzo-p-dioxin (TCDD)[9-11] in fish was the application of relatively high concentrations of these chemicals in the ambient water. The concentrations exceeded the aqueous solubility by several orders of magnitude. Although aqueous solubility lies between 74 and 400 pg/L,[17,26-28] several authors conducted experiments with concentrations $>10^4$ pg/L.[12,16,17] Because only the "truly" dissolved chemical is able to be absorbed by the fish via gills,[20,21] the use of supersaturated concentrations (dissolved plus sorbed) will clearly underestimate BCF values.[17] Low uptake of superlipophilic chemicals is caused by low bioavailability rather than by low bioconcentration potentials itself.

Fifth, using several different approaches, we have presented evidence that the thesis concerning bioconcentration of superlipophilic chemicals in aquatic animals from ambient water has to be revised, at least in parts. The thesis that chemicals with log $K_{OW} > 6$ are bioconcentrated to a significantly smaller degree than predicted from their K_{OW} is not valid any longer as a general statement. However, this statement is still valid for several superlipophilic chemicals with peculiar molecular size or shape, such as paraffins, organosilicon compounds with a very long chain, and organic colorants.[37] These chemicals possess a low membrane permeability. Actually, the threshold of membrane permeability does not apply any longer. It clearly lies above the cross section of 9.5 Å. Moser and Anliker[37] have chosen this molecular size parameter >10.5 Å where bioconcentration tests with fish are not necessary.

Finally, with increasing lipophilicity the uptake velocity is clearly declining and steady-state conditions are not achieved within a few weeks, but in many instances only after several or even after many months. As one consequence, BCF values of superlipophilic chemicals have to be evaluated under flow-through conditions rather than batch conditions. It appears to be self-evident that aquatic organisms should be exposed only to ambient concentrations below the water solubility. However, to fulfill both experimental conditions with superlipophilic compounds, severe practical problems emerge.

ACKNOWLEDGMENT

The authors are grateful to Drs. O. Hutzinger, D.W. Hawker, L. Birnbaum, W. Butte, H. Fiedler, H. Beck, D. Sijm, F. Gobas, J. Altschuh, L.S. McCarthy, and P. Cook for helpful suggestions, for providing data, and for discussions.

REFERENCES

1. Ernst, W. "Accumulation in Aquatic Organisms," in *Appraisal of Tests to Predict the Environmental Behaviour of Chemicals*, P. Sheehan, F. Korte, W. Klein, and Ph. Bourdeau, Eds. (New York: John Wiley & Sons, 1985), Chapter 4.4, pp. 243–255.
2. Spacic, A. and J.L. Hamelink. "Bioaccumulation," in *Fundamentals of Aquatic Toxicology: Methods and Applications*. G.M. Rand and S.R. Petrocelli, Eds. (Washington, DC: Hemisphere Publishing Corporation, McGraw-Hill International Book Co., 1985), Chapter 13, pp. 495–525.
3. Macek, K.J., S.R. Petrocelli, and B.H. Sleight, III. "Considerations in Assessing the Potential for, and Significance of Biomagnification of Chemical Residues in Aquatic Food Chains," in *Aquatic Toxicology*, L.L. Marking, R.A. Kimmerle, Eds., ASTM STP 667, (Philadelphia, PA: ASTM, 1979), pp. 251–268.
4. Connell, D.W. "Biomagnification by Aquatic Organisms — A Proposal," *Chemosphere* 19:1573–1587 (1986).
5. Bruggeman, W.A., A. Opperhuizen, A. Wijbenga, and O. Hutzinger. "Bioaccumulation of Super-Lipophilic Chemicals in Fish," *Environ. Toxicol. Chem.* 7:173–189 (1984).
6. Gobas, F.A.P.C., A. Opperhuizen, and O. Hutzinger. "Bioconcentration of Hydrophobic Chemicals in Fish: Relationship with Membrane Permeation." *Environ. Toxicol. Chem.* 5:637–646 (1986).
7. Gobas, F.A.P.C., K.E. Clark, W.Y. Shiu, and D. MacKay. "Bioconcentration of Polybrominated Benzenes and Biphenyls and Related Superhydrophobic Chemicals in Fish: Role of Bioavailability and Elimination into the Feces," *Environ. Toxicol. Chem.* 8:231–245 (1989).
8. Connell, D.W. and D.W. Hawker. "Use of Polynominal Expressions to Describe the Bioconcentration of Hydrophobic Chemicals by Fish," *Ecotoxicol. Environ. Saf.* 16:242–257 (1988).
9. Mehrle, P.M., D.R. Buckler, E.E. Little, L.M. Smith, J.D. Petty, P.H. Peterman, D.J. Stalling, G.M. De Graeve, J.J. Coyle, and W.J. Adams. "Toxicity and Bioconcentration of 2,3,7,8-Tetrachlorodibenzodioxin and 2,3,7,8-Tetrachlorodibenzofuran in Rainbow Trout," *Environ. Toxicol. Chem.* 7:47–62 (1988).
10. Branson, D.R., I.T. Takahashi, W.M. Parker, and G.E. Blau. "Bioconcentration of 2,3,7,8-Tetrachlorodipenzo-p-dioxin in Rainbow Trout," *Environ. Toxicol. Chem.* 4:779–788 (1985).
11. Cook, P.M., D.W. Kuehl, M.K. Walker, and R.E. Peterson. "Bioaccumulation and Toxicity of TCDD and Related Compounds in Aquatic Ecosystems," in *Banbury Report 35: Biological Basis for Risk Assessment of Dioxins and Related Compounds* (Cold Spring Harbor, New York: Cold Spring Harbor Laboratory Press, 1991), pp. 143–167.

12. Opperhiuzen, A., W.E.V.D. Velde, F.A.P.C. Gobas, D.A.K. Liem, J.M.D.V.D. Steen, and O. Hutzinger. "Relationship Between Bioconcentration in Fish and Steric Factors of Hydrophobic Chemicals," *Chemosphere* 14:1871–1896 (1985).

13. Miyata, H., K. Takayama, J. Ogaki, M. Mimura, and T. Kashimoto. "Study on Polychlorinated Dibenzo-p-Dioxins and Dibenzofurans in Rivers and Estuaries in Osaka Bay in Japan," *Toxicol. Environ. Chem.* 7:91–101 (1988).

14. Oehme, M., S. Manø, E. Brevik, and J. Knutzen. "Determination of Polychlorinated Dibenzofuran (PCDF) and Dibenzo-p-Dioxin (PCDD) Levels and Isomer Pattern in Fish, Crustacea, Mussels and Sediment Samples from a Fjord Region Polluted by Mg-Production," *Fresenius J. Anal. Chem.* 335:987–997 (1989).

15. Zacharewski, T., L. Safe, S. Safe, B. Chittim, D. DeVault, K. Wiberg, P.-A. Bergqvist, and C. Rappe. "Comparative Analysis of Polychlorinated Dibenzo-p-Dioxin and Dibenzofuran Congeners in Great Lakes Fish Extracts by Gas Chromatography-Mass Spectrometry and in vitro Enzyme Induction Activities," *Environ. Sci. Technol.* 232:730–735 (1989).

16. Muir, D.C.G., A.L. Yarechewski, A. Knoll, and G.R.B. Webster. "Bioconcentration and Disposition of 1,3,6,8-Tetrachlorodibenzo-p-Dioxin and Octachlorodibenzo-p-Dioxin by Rainbow Trout and Fathead Minnows," *Environ. Toxicol. Chem.* 5:261–272 (1986).

17. Gobas, F.A.P.C. and S.M. Schrap. "Bioaccumulation of Some Polychlorinated Dibenzo-p-Dioxins and Octachlorodibenzofuran in the Guppy (*Poeciia reticulata*)," *Chemosphere* 20:495–512 (1990).

18. Muir, D.C.G., W.K. Marschall, and G.R.B. Webster. "Bioconcentration of PCDDs by Fish: Effects of Molecular Structure and Water Chemistry," *Chemosphere* 14:829–833 (1985).

19. Miyata, H., K. Takayama, M. Mimura, and T. Kashimoto. "Investigation on Contamination Sources of PCDDs and PCDFs in Blue Mussel," *Chemosphere* 19:517–520 (1989).

20. Servos, M.R. and D.C.G. Muir. "Effect of Dissolved Organic Matter from Canadian Shield Lakes on the Bioavailability of 1,3,6,8-Tetrachlorodibenzo-p-Dioxin on the Amphipod *Crangonyx laurentianus*," *Environ. Toxicol. Chem.* 8:141–150 (1989).

21. Black, M.C. and J.F. McMcarthy. "Dissolved Organic Macromolecules Reduce the Uptake of Hydrophobic Organic Contaminants by the Gills of Rainbow Trout (*Salmo gairdneri*)," *Environ. Toxicol. Chem.* 7:593–600 (1988).

22. Neely, W.B. "Estimating Rate Constants for Uptake and Elimination of Chemicals by Fish," *Environ. Sci. Technol.* 13:1506–1510 (1979).

23. Opperhuizen, A. and D.T.H.M. Sijm. "Bioaccumulation and Biotransformation of Polychlorinated Dibenzo-p-Dioxins and Dibenzofurans in Fish," *Environ. Toxicol. Chem.* 9:175–186 (1990).

24. Geyer, H., I. Scheunert, D.C.G. Muir, and A. Kettrup. "Comparison of the Octachlorodibenzo-p-Dioxin (OCDD) Bioaccumulation Potential in Aquatic Biota Estimated by Different Methods," in *Organohalogen Compounds*," Vol. 1, O. Hutzinger and H. Fiedler, Eds. Dioxin '90 — EPRI-Seminar Toxicology, Environment, Food, Exposure-Risk. (Bayreuth: Verlag Ecoinforma, 1990), pp. 341–346.

25. Geyer, H.J., D.C.G. Muir, I. Scheunert, and A. Kettrup. "Bioconcentration of Octachlorodibenzo-p-Dioxin (OCDD) in Mussels and Fish," 11th International Symposium on Chlorinated Dioxin and Related Compounds. Research Triangle Park, NC, (1991).

26. Geyer, H.J., I. Scheunert, and F. Korte. "Relationship Between the Lipid Content of Fish and Their Bioconcentration Potential of 1,2,4-Trichlorobenzene," *Chemosphere* 14:545–555 (1985).

27. Shiu, W.Y., W. Doucette, F.A. Gobas, A. Andreu, and D. Mackay. "Physical-Chemical Properties of Chlorinated Dibenzo-p-dioxins," *Environ. Sci. Technol.* 22:651–658 (1988).

28. Doucette, W.J. and A.W. Andreu. "Aqueous Solubility of Selected Biphenyl, Furan, and Dioxin Congeners," *Chemosphere* 17:243–252 (1988).

29. Geyer, H.J., P. Sheehan, D. Kotzias, D. Freitag, and F. Korte. "Prediction of Ecotoxicological Behaviour of Chemicals: Relationship Between Physio-chemical Properties and Bioaccumulation of Organic Chemicals in the Mussel *Mytilus edulis*," *Chemosphere* 11:1131–1134 (1982).

30. Veith, D.G., D.L. DeFoe, and B.V. Bergstedt. "Measuring and Estimating the Bioconcentration Factor of Chemicals in Fish," *J. Fish. Res. Board. Can.* 36:1040–1048 (1979).

31. Huckins, J.N., D.L. Stalling, J.D. Petty, D.R. Buckler, and B.T. Johnson. "Fate of Kepone and Mirex in the Aquatic Environment," *J. Agric. Food. Chem.* 30:1020–1027 (1982).

32. Clark, T., D. Clark, S. Paterson, D. Mackay, and R.J. Norstrom. "Wildlife Monitoring, Modeling and Fugacity," *Environ. Sci. Technol.* 22:120–127 (1988).

33. Connolly, J.P. and C.J. Pedersen. "A Thermodynamic-Based Evaluation of Organic Chemical Accumulation in Aquatic Organisms," *Environ. Sci. Technol.* 22:99–103 (1988).

34. Kulka, M. Canadian Patent 702,144, 1965; *Chem. Abstr.* 62:16261a,b (1965).

35. Geyer, H.J., I. Scheunert, R. Brüggemann, Ch. Steinberg; F. Korte, and A. Kettrup. "QSAR for Organic Chemical Bioconcentration in *Daphnia*, Algae, and Mussels," *Sci. Total Environ.* 109/110:387–394 (1991).

36. Butte, W. "Mathematical Description of Uptake, Accumulation and Elimination of Xenobiotics in a Fish/Water System," in *Bioaccumulation in Aquatic Systems; Contributions to the Assessment; Proceedings of an International Workshop,* 1st ed., R. Nagel and R. Loskill, Eds. (Weinheim: VCH Publishers, 1991), pp. 29–42.

37. Moser, P. and R. Anliker. "BCF and P: Limitations of the Determination Methods and Interpretation of Data in the Case of Organic colorants," in *Bioaccumulation in Aquatic Systems; Contributions to the Assessment; Proceedings of an International Workshop* 1st ed., R. Nagel and R. Loskill, Eds. (Weinheim: VCH Publishers, 1991), pp. 13–28.

38. Burkhard, L.P. and D.W. Kuehl. "N-Octanol/Water Partition Coefficients by Reverse Phase Liquid Chromatography/Mass Spectrometry for Eight Tetrachlorinated Planar Molecules." *Chemosphere* 15:163–167 (1986).

39. Rippen, G. *Handbuch Umweltchemikalien* (Landsberg: ECOMED, 1990).

40. Broman, D., Näf, C., Rolff, C., and Zebühr, Y. "Occurence and Dynamics of Polychlorinated Dibenzo-p-dioxins and Dibenzofurans and Polycyclic Aromatic Hydrocarbons in the Mixed Surface Layer of Remote Coastal and Offshore Waters of the Baltic," *Environ. Sci. Technol.* 25(11):1850–1864 (1991).
41. Geyer, H.J., Scheunert, I., Korte, F. "Correlation Between the Bioconcentration Potential of Organic Environmental Chemicals in Humans and their *N*-Octanol/Water Partition Coefficients," *Chemosphere* 16:239–252 (1987).
42. Luckas, B and Oehme, M., "Characteristic Contamination Levels for Polyclorinated Hydrocarbons, Dibenzofurans and Dibenzo-*p*-dioxins in Bream (*Abramis Brama*) From the River Elbe" *Chemosphere*, 21 (1–2):79–89 (1990).

CHAPTER **19**

Behavior of Soil Microflora in Pesticide Degradation

J.-C. Fournier, C. Catroux, M.-P. Charnay, and Gunalan

I. INTRODUCTION

Certain main conditions for an efficient microbial degradation of pesticides in soil are presented. The metabolic behavior of the pesticide degrading microflora influences greatly the persistence of the compound both in agronomic and natural environmental situations. The results of some experiments which illustrate this are also presented. The conditions of these experiments were described in detail in another publication.[1] Small samples of soil (50 g) sieved using the 0–3-mm fraction were treated with ^{14}C-labeled pesticides. The soil samples were placed in closed glass jars of 2 L and incubated at 20°C in the dark. The $^{14}CO_2$ released from the treated samples was trapped in a NaOH solution, and the radioactivity was estimated by liquid scintillation counting. Microorganisms able to use 2,4-dichlorophenoxyacetic acid (2,4-D) or carbofuran as the sole carbon source were enumerated with the "most probable number" method.

II. MAIN REQUIREMENTS FOR THE MICROBIAL DEGRADATION OF PESTICIDES IN SOIL

The soil microflora can degrade most of the pesticides we actually know. However, the microbial pesticide degradation in soil is not systematically very

0-87371-616-7/93/$0.00 + $.50

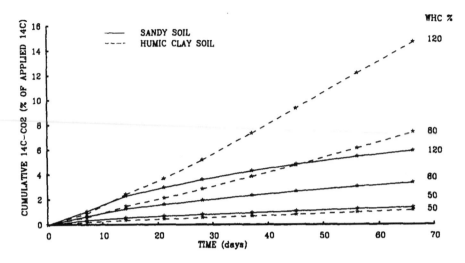

Figure 1. Effect of the soil humidity on the degradation of atrazin ([^{14}C]ethyl) at 3 mg/kg in a humic clay soil (Bressey) and a sandy soil (Auxonne).

efficient and does not always represent the prevalent phenomenon for each pesticide transformation.

Three main requisites allow the microbial degradation of pesticides in soil:

A. Potentially Degrading Strains Must Be Present

This seems evident, but in fact a specific and efficient degrading strain present in some soils, areas, or environmental conditions can lack in other circumstances. Even if a degrading organism is present, its distribution in the soil can be very heterogeneous; furthermore, it can be very scarce: sometimes 10 or less microorganisms per gram of soil.

B. The Microbial Activity and Biological Availability of the Pesticide Depend on the Physical and Chemical Conditions of the Environment

Certain effects of this type can explain the relation between humidity and degradation of atrazin in two soils (Figure 1). For this experiment, different samples of a sandy soil and a clay humic soil were moistened to 50, 80, or 120% of their water-holding capacity. At the lower humidity, the atrazin degradation occurred very slowly in the two soils. With an increase in soil humidity, the rate of degradation increased. However, the maximum degradation was observed in the organic clay soil. Therefore, it seems that in this experiment, maximum microbial biomass of the organic soil was the main factor acting on the atrazin degradation. Higher availability of the herbicide in the sandy soil was not sufficient to allow a rapid degradation of the compound.

Table 1. Different Microbial Strategies for the Degradation of Pesticides in Soil

Usual term	Metabolism	Cometabolism
Main support of growth	Pesticide	Other substrates
Number of pesticides concerned	Few	Most
Carbon and Energy Resources	Low amounts but specific	Sometimes great amounts but high competition with surrounding microorganisms
Microflora Implicated	A few specific bacterial strains	Often omnivorous organisms as fungi
Main agricultural problem	Soil adaptation to degradation after microbial enrichment	Accidental excessive persistence
Main environmental problem	Mobility	Excessive pesticide or metabolite persistence and mobility

C. Pesticide Degrading Microorganisms Have to Find Energy and Carbon Sources for Survival and Growth

These carbon sources can be the pesticide itself: in this case the term of "metabolism" is used simply to identify this "microbial strategy" of degradation. However, in many cases the source of carbon and energy cannot be the pesticide itself but other soil organic substrates; thus, the term of "cometabolism" is generally used.[2]

These two strategies of the pesticide degrading microflora are of great interest for the understanding of pesticide behavior. Some essential characteristics of each one are summarized in Table 1. Relatively few pesticides can be used by soil microbial strains as the sole or main carbon source. To do this the microorganisms must have the capability of achieving a long sequence of metabolic transformations including steps that provide energy. Therefore, it is not surprising that only a small number of microorganisms, generally bacteria, have been isolated and described. The ability of many of these bacteria to store part of their genetic information on different transferable elements such as plasmids,[3,4] could explain partly their development and efficiency in the environment.

In contrast to the previous observations, the cometabolic degradation concerns most pesticides. Cometabolic microorganisms are unable to achieve a complete breakdown of the pesticide. The lack or inhibition of certain enzymes causes the degrading strains to use a cosubstrate for growth. In soil conditions these cosubstrates must be found among endogeneous or sometimes soil-introduced organic matter. Consequently, the cometabolic organisms have to compete efficiently for these cosubstrates with the surrounding microorganisms. It seems in these conditions that omnivorous microorganisms such as fungi could be more specially effective.

We can also say that theoretically the cometabolic transformation of a substrate by a strain growing in axenic conditions leads to the accumulation of metabolites. Fortunately, in soil several microorganisms or sometimes abiotic and biotic systems generally coexist to obtain a more complete breakdown of metabolites.

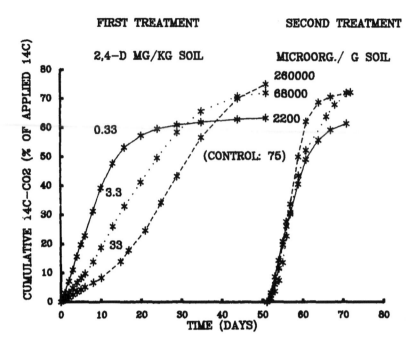

Figure 2. Effect of concentration and repetition of treatments on soil enrichment in pesticide degrading microorganisms and biodegradation of [$^{1\text{-}14}$C]2,4-D.

III. AGRONOMIC AND ENVIRONMENTAL CONSEQUENCES OF THE DIFFERENT MICROBIAL STRATEGIES FOR PESTICIDE DEGRADATION

The metabolic and cometabolic behavior of the degrading microflora have both favorable and unfavorable consequences on pesticide use. Two main problems of the agronomic and the environmental behavior of pesticides will be discussed here.

A. Decreasing Soil Persistence of Some Pesticides after Repeated Treatments

This important agronomic problem is related to the loss in persistence of some pesticides after several years of repeated treatments in fields. It is related to a "metabolic" process of microbial degradation.

An example of soil adaptation to the herbicide 2,4-D is shown in Figure 2. After the first treatment with different doses of 2,4-D, we observed an almost proportional increase in the degrading population size. The number of microorganisms able to use 2,4-D as the sole carbon source varied from 10 microorganisms per gram of soil to more than 200,000 per gram after degradation of

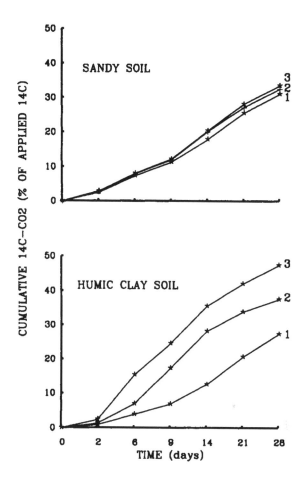

Figure 3. Effect of three successive treatments on the degradation of carbofuran ([^{14}C]carbonyl) at 3 mg/kg in a sandy soil (Auxonne) and a humic clay soil (Bressey).

the highest dose of pesticide. In these conditions, a second treatment with the same dose of pesticide could be degraded quickly.

Because 2,4-D is not used for soil treatments, the previous example is only interesting as a model study. This is not the case for the following experiment which concerned degradation of the soil insecticide carbofuran (Figure 3). Its main purpose was to compare two soils to develop their capability for degrading the pesticide. We observed that an adaptation to the carbofuran degradation only appeared in a humic soil, but not in a sandy soil with a low organic level. The enumeration of degrading microorganisms confirmed their initial presence and growth only in the organic soil.

Many experimental and theoretical questions might be considered. Among these, the small sample size can sometimes reduce the probability of retaining

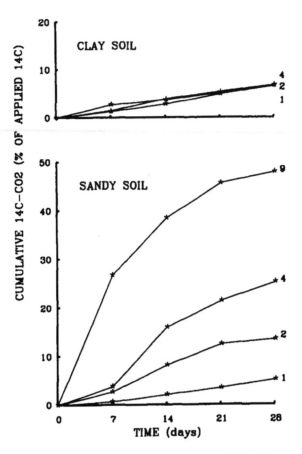

Figure 4. Effect of repeated treatments on the degradation of CIPC (^{14}C ring) at 3.2 mg/kg in a sandy soil (Auxonne) and a clay soil (Veuvey).

at least one degrading organism. This may involve differences between results from laboratory and in situ experimentations. Second, we do not consider whether new strains genetically recombined could naturally and quickly emerge in soil conditions.

Figure 4 provides a second example of differential adaptation of two soils. In contrast with the previous experiment, the quicker degradation of the herbicide CIPC was obtained in a sandy soil. This observation could be related to the low microbial availability of the pesticide in the organic soil. Adsorption or fixation on the organic matter may limit the concentration of the compound in the soil solution and reduce the possibility of microbial growth.

B. Excessive Persistence of Pesticides in the Environment

With the exception of accidents, the excessive persistence of pesticides or biocidal metabolites is unacceptable, and damage would lead to discarding a compound for further agronomic use. However, in many cases the absence of damage does not mean a real or a whole degradation of the pesticide. The compound can only move in the soil, to be diluted or transferred out of the cultivated areas.

Excessive persistence in the environment occurs more frequently, but not systematically with pesticides degraded by cometabolic processes. Even if these compounds are weakly soluble, they can move slowly to the deep layers of the soil, that is to say, in soil areas where the organic content and size of the microbial biomass are very limited. This is illustrated by the results presented in Figure 5. In this laboratory experiment, we used small soil samples from different depths. Then we studied the degradation of labeled 2,4-D and atrazin in optimal conditions of temperature, moisture, and oxygenation. 2,4-D was degraded more quickly in top soil than in subsoil samples. However, we observed that at least a few microorganisms were always present in the different subsoil samples and that these organisms could proliferate using 2,4-D as the main carbon source. This explains that the pesticide was generally degraded after a relatively short time of incubation.

Microbial degrading ability to 2,4-D seemed less evident when the concentration of the pesticide in the subsoil samples were too low. The unexpected slow degradation of certain compounds, in soil or water, have been previously reported.[4,5] In the example presented in Figure 6, we observed that the half-life of 2,4-D increased in the subsoil samples when doses of applied pesticide decreased. This observation seems related to the difficulty in degrading microorganisms to proliferate in this situation of limited carbon availability.

In contrast with 2,4-D, atrazin is only degraded by cometabolic processes, and no specific population can use it for growth. Furthermore, the low levels of organic carbon in the subsoil cannot sustain the growth of a large cometabolic microflora and allow a rapid degradation of the herbicide (Figure 5).

The following experiment (Figure 7) conducted with top soil samples showed an increase in the degradation rate of atrazin after the addition of large quantities of glucose, cellulose, and nitrogen in a silty clay soil with a medium microbial biomass. The addition of glucose had a marked but short stimulating effect on the atrazin degradation. In the same soil, cellulose addition also resulted in an increasing degradation rate of the herbicide, but only after a small lag phase. In contrast with the glucose effect, the cellulose effect persisted for over a month. We did not observe the cellulose effect in the second experimental soil.

IV. CONCLUSION

The few experimental examples presented in this chapter only provide an imperfect view of the complex behavior of pesticide degrading populations in soil. Many aspects and many specific conditions should still be investigated. It is evident that we are not able to estimate the degrading efficiency of soil microflora from physicochemical and pedological data alone. This means that

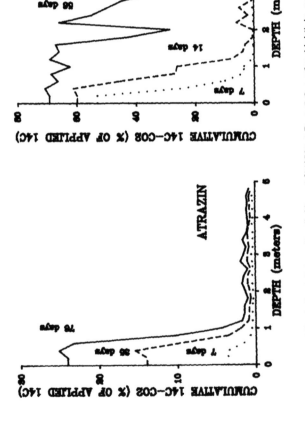

Figure 5. Degradation of atrazin ([[1-14C]ethyl] at 0.1 mg/kg (left) and [1-14C]2,4-D at 3.2 mg/kg (right) in samples from different depths (laboratory experiments).

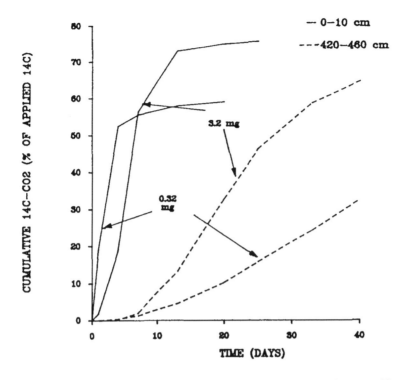

Figure 6. Degradation of [1-14C]2,4-D at 3.2 and 0.32 mg soil in samples from two different depths.

microbiologists must define their own experimental conditions and criteria of study in a better way. Problems of microbial adaptation to degradation of pesticides after repeated field treatments and determination of the persistence of compounds in other environments (subsoil, groundwater, etc.) must be studied with special care. It is difficult to solve these problems using only the limited number of soils often recommended for European coordinated studies and regulation requirements.

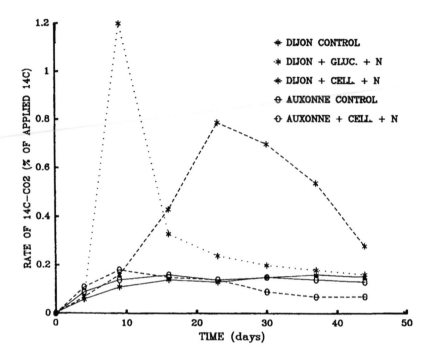

Figure 7. Effect of glucose or cellulose at 5 g/kg of soil and ammonium sulfate at 1 g/kg soil on the degradation rate of atrazin ([^{14}C]ethyl) at 3.2 mg/kg in a silty clay soil (Dijon) and a sandy soil (Auxonne).

REFERENCES

1. Fournier, J.C. "Aspects du comportement de la microflore degradant les produits phytosanitaires dans le sol," Thesis, University of Perpignan, Dijon (Laboratory of Soil Microbiology) (1989).
2. Alexander, M. "Role of Cometabolism," in *Microbial Degradation of Pollutants in Marine Environments*, A.V. Bourquin and P.H. Pritchard, Eds. U.S. EPA, Gulf Breeze, FL, Publ. EPA-600/9-79-012, (1979) pp. 67–75.
3. Chaudhry, G. and G.H. Huang. "Isolation and Characterization of a New Plasmid from a *Flavobacterium* sp. Which Carries the Genes for Degradation of 2,4-Dichlorophenoxyacetate," *J. Bacteriol.* 170:1897–1902 (1988).
4. Pemberton, J.M. "Degradative Plasmids," *Int. Rev. Cytol.* 84:155–183 (1983).
5. Boethling, R.S. and M. Alexander. "Effect of Concentration of Organic Chemicals on Their Biodegradation by Natural Microbial Communities," *Appl. Environ. Microbiol.* 37:1211–1216 (1979a).
6. Boethling, R.S. and M. Alexander. "Microbial Degradation of Organic Compounds at Trace Levels," *Environ. Sci. Technol.* 13:989–991 (1979b).

CHAPTER 20

Modeling the Uptake of Organic Compounds into Plants

S. Trapp

ABSTRACT

A fugacity model for the uptake of chemicals into plants was developed. Laboratory experiments were used to validate the simulation model. The plant-soil bioconcentration factor depends on the ratio of K_{ow}:K_{oc} values of the chemical, the lipid fraction of the plant, the organic carbon content of the soil, the water contents, and the transfer and metabolism kinetics.

I. INTRODUCTION

Plants may be contaminated by hazardous chemicals in numerous ways, e.g., by the use of pesticides, the deposition of chemicals from air, the agricultural use of sewage sludge, the dumping of toxic wastes, or the contamination of ground- and soil water.

It is thus not surprising that toxic residues have been found in plants. Reported concentrations are high (see Table 1) and in the range of the sediment concentrations of polluted rivers. Uptake into plants is the first step for an accumulation in the terrestrial food web and should be considered further.

0-87371-616-7/93/$0.00 + $.50

Table 1. Concentrations of Environmental Chemicals in Pine Needles Compared to Water and Sediment of the Elbe River Close to Hamburg

Substance	Water (μg/L)	Sediment (μg/kg)	Pine needles (μg/kg)
α-HCH	0.023	ca. 1, max 129	5–12
Lindane	0.021	ca. 3, max 132	7–22
HCB	0.020	ca. 20, max 1180	1–3
p,p'-DDE	0–0.004	ca. 5	0.8–2.5
p,p'-DDT	0–0.004	ca. 10	2–7
PCB[a]	0.015	200–800	10–300

Source: ARGE Elbe[1] and Reischl.[2]

[a] Aroclor 1260 rsp. Chlophen A60.

Table 2. Properties of the Investigated Chemicals

Name	log K_{ow}	log K_{oc}	H	MW
Atrazine	2.71	2.33	8.05×10^{-9}	215.7
TriCB	3.98	3.35	0.17	181.5
TetraCB	4.65	3.84	0.1	215.8
Dieldrin	5.48	4.44	4.4×10^{-4}	380.9
PCB	5.92	4.75	0.0076	326.4
HCB	5.47	4.43	0.054	284.8
DDT	6.19	5.14	2.14×10^{-3}	354.5

Note: Log K_{ow} is the logarithm of the partition coefficient between n-octanol and water, log K_{oc} is the logarithm of the partition coefficient between soil organic carbon and water, H is the dimensionless partition coefficient between air and water, and MW is the molecular weight in grams per mole. TriCB = 1, 2, 4-trichlorobenzene; TetraCB = 1, 2, 3, 5-tetrachlorobenzene; PCB = 2, 2', 4, 4', 6-pentachlorobiphenyl; HCB = hexachlorobenzene.

II. LABORATORY EXPERIMENTS

Laboratory experiments were conducted by Scheunert and colleagues.[3] Test soil contaminated with ^{14}C-labeled chemicals was filled in an exsiccator. Barley plants were seeded and grown for one week. Afterward, concentrations in soils, plants, and air were measured. Radioactivity was balanced. Additionally, the fraction of metabolites was determined. Among the investigated chemicals were atrazine, chlorobenzenes, the insecticides dichlorodiphenyltrichloroethane (DDT) and dieldrin, hexachlorobenzene (HCB), and polychlorobiphenyls (PCBs) (Table 2).

A. Results of the Laboratory Experiments

The distribution of ^{14}C in the system was measured.[3,4] The accumulation into plants is here expressed as plant:soil concentration ratio of the measured total ^{14}C, including metabolites (plants fresh weight). The bioconcentration factors

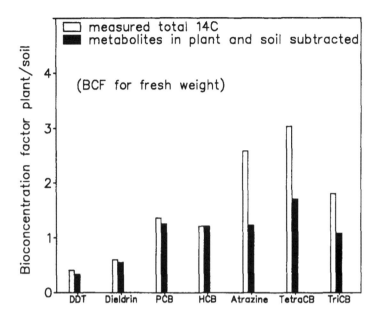

Figure 1. Measured plant/soil bioconcentration factor.

for the parent compounds are obtained when the fraction of metabolites in soil and plant are subtracted from the total ^{14}C activities in soil and plant. This produces for the more polar compounds a large difference (Figure 1). The plant/soil bioconcentration factor is surprisingly rather constant for most chemicals. Only DDT and dieldrin show lower uptake into the barley plants.

III. FORMULATION OF THE MODEL

For the calculation of the chemical behavior in the soil-air-plant system, the following aspects have to be considered:

- equilibrium conditions
- transfer rates
- metabolism rates (here: measured)
- growth of the plants

The system is divided into four compartments, namely, soil, air, roots, and shoots. These compartments are coupled by transfers.

A. Calculation of the Equilibrium

The equilibrium constant soil matrix to water is the well-known K_d value and may be calculated using the organic carbon content, the K_{oc}, the pore water, and the air-filled pores. The equilibrium constant air to water is the dimensionless Henry's law constant.

The partitioning of chemicals into plants scarcely has been investigated. Recent investigations have shown that the uptake of chemicals into plants is related to the water and lipid content.[4-7] A simplified formulation is:

$$K_p = f \cdot a \cdot K_{ow}^b + W_p$$

where K_p = concentration ratio plant to water (in equilibrium)
       f = lipid content of the plant
      W_p = water content of the plant

Plant lipids may vary in their properties and differ from octanol. The parameters a and b correct this, if necessary. The density of the plant is assumed to be equal to that of water and does not appear in the equation.

Over a long range, the equilibrium is independent of the K_{ow} and determined by the water content of the plant. More lipophilic chemicals ($K_{ow} > 100$) tend to sorb to the lipid fraction of the plant, and the concentration ratio plant to water is increasing for higher K_{ow}-values.

The plant-to-soil bioconcentration factor (BCF) is the ratio of the K_p to the K_d value (for equilibrium conditions). For water soluble substances the BCF is determined by the ratio of water content in plant and soil. For example, when a plant has a water content of 80% and the soil moisture is 20%, the BCF would be 4.

For hydrophobic chemicals, the plant/soil equilibrium is mainly determined by the partitioning between the lipid fraction in the plant and the soil organic carbon fraction. It is a function of the ratio $K_{ow}:K_{oc}$. Lipid content and organic carbon content describe the soil and plant properties. The K_{oc} is correlated to the K_{ow}. There exist a number of regression equations that differ considerably (Figure 2). Here the equation of Schwarzenbach is used.[8]

The BCF plant to soil as a function of the K_{ow} is rather constant. For low K_{ow} values, concentration in the plant is higher than in the soil due to higher water content. The curve has a minimum in the medium range of K_{ow} and then begins to increase again slowly. This equilibrium function fits the results very well in the medium range. The BCF of hydrophobic chemicals tends to be overestimated (Figure 3). Herefore, two major reasons may be found:

1.   For hydrophobic chemicals it is difficult to estimate the K_{oc} from the K_{ow} (measured values should be used when available).
2.   Hydrophobic chemicals are strongly sorbed to the soil matrix and are only slowly transferred into air or plant. The equilibrium is not achieved. Subsequently, transfer rates have to be taken into account.

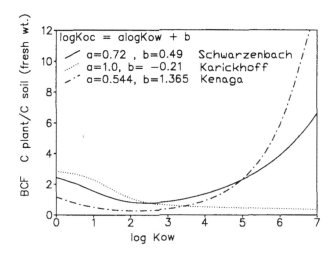

Figure 2. Dependence of the calculated BCF from the K_{oc}-regression (here: roots to soil).

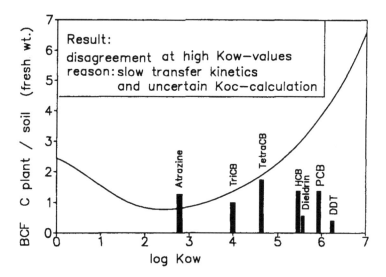

Figure 3. Plant/soil bioconcentration factor (fresh weight); bars: measured; lines: calculated.

B. Uptake Pathways of Chemicals into Plants

A chemical can move in the soil by diffusion. It may diffuse into deeper soil layers, into roots, or into air. Water soluble chemicals are taken up by the plant with the water it needs. While the roots are growing in the soil, they may have contact with contaminants; and the contaminants may sorb to the roots.

1. Diffusion Soil to Root

The diffusion from soil into roots may be described as the diffusion into a cylinder. It occurs in air-filled pores; diffusion in water pores is very slow and thus negligible.

$$N = D_{eff} \cdot 2 \cdot \pi \cdot L/[\ln (R2/R1)] (C_s \cdot H/K_d - C_p \cdot H/K_p)$$

where D_{eff} = effective diffusivity in air-filled pores (m²/sec)
        N = flux of chemical (kg/sec)
        L = root length (m)
        C = concentration (s = soil, p = plant) (kg/m³)
        H = dimensionless Henry's law constant
    H/K = equilibrium partitioning to air.

The parameter R1 is the radius of the root. R2 describes the diffusion length and is difficult to estimate: at the time of contamination R2 should be close to R1, because the root is in direct contact with the contaminated soil. This means uptake is rapid at first. Then, a deficiency zone of chemical develops in the vicinity of the root. The radius of this zone was found to be 1.5 mm for phosphorus.[9] The situation is complicated furthermore because the root may grow in length as well as thickness.

2. Uptake with Water

Water-soluble chemicals enter the plant with the soil water taken up by roots to form the transpiration stream. The formulation of the process is:

$$N = T \cdot (C_s/K_d)$$

where T is the flux of water (m³/sec).

Further translocation with the transpiration stream into higher parts of the plant depends on the K_{ow}. Very hydrophilic and very hydrophobic chemicals are slowly translocated in the plant. Transport in the xylem occurs mainly for medium hydrophobic chemicals (Figure 4) and is rapid, because a plant transpires large amounts of water. Most herbicides are in this range. The concentration ratio transpiration stream to soil solution can be calculated:[10]

$$TSCF = 0.784 \cdot \exp[-(\log K_{ow} - 1.78)^2/2.44]$$

C. Uptake of Chemicals by Aerial Plant Parts

When the partial pressure of a chemical is not too low it will volatilize from the soil into the air and then enter aerial plant parts:

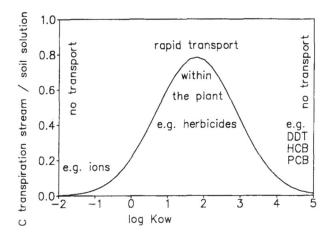

Figure 4. Transpiration stream concentration factor.

$$N = Tk \cdot A \cdot (C_a - C_p \cdot H/K_p)$$

where Tk = transfer coefficient representing the conductance air to leave (m/sec)

A = foliage area (m^2)

C_a = air concentration (kg/m^3)

Reported values for Tk differ: Thompson[11] makes the assumption that the cuticula of the leave is a perfect sink for the chemical (its resistance to transfer is zero). The assumption is valid for sorption on the cuticula. He estimates Tk to 0.05 m/sec for pesticides with a molecular weight of 300 g/mol.

For an incorporation into the plant or for a transport from the plant into air, two other conductances (acting parallel) must be considered: stomata and cuticular conductance. Values for Tk ranging from 2.5×10^{-3} m/sec to 0.125×10^{-5} m/sec were reported when measured cuticular and stomata resistances were added (species: *Citrus aurantium*).[5]

IV. COMPARISON OF MODEL AND EXPERIMENTAL RESULTS

Transfer kinetics together with equilibrium constants were used for a time-dynamic simulation of the soil-air-plant system.[4] Only the parent compounds are comparable. The results are in good agreement (Figure 5).

The model allows an interpretation of the measured results and clarifies the processes in the system. This is particularly possible by looking at the fugacities in the system. Fugacity is the escaping tendency of a chemical (with the dimension of pressure). Diffusion always occurs from the phase with the higher

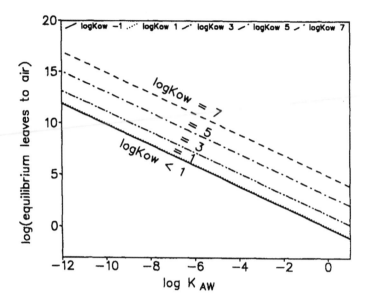

Figure 5. Equilibrium between leaves and air.

fugacity to the phase with the lower one. Two phases are in equilibrium when the fugacity is equal.

First, it must be mentioned that the laboratory system was very small (0.01 m³); air concentrations were high (due to a low gas exchange rate); and plants were seeded, grown for only 1 week, and then harvested (when they were still very small). Equilibrium or steady-state conditions are reached much faster than in a large system (e.g., in a field).

Chemicals that have similar properties (water solubility, K_{ow}, vapor pressure, molecular weight, metabolization rates) behave comparably. Typical behavior profiles are:

1. Behavior of hydrophobic chemicals with a high vapor pressure (e.g., HCB, PCBs, chlorobenzenes, etc.): These chemicals enter roots via diffusion and foliage via air. The concentration in aerial plant parts is strongly influenced from the concentration in air.

2. Chemicals with medium polarity and low vapor pressure (e.g., atrazine, most herbicides): These chemicals are transported with the transpiration stream and tend to accumulate in the leaves. The concentration in aerial plant parts is only weakly influenced by air concentration (except when the source is in air).

3. Hydrophobic chemicals with low vapor pressure (e.g., DDT, dioxins, etc.) (Figure 7): DDT shows again a different behavior. Here also, roots are in equilibrium with the soil, and shoots are close to equilibrium with air. Due to the low vapor pressure of DDT, however, it volatilizes slowly. The air

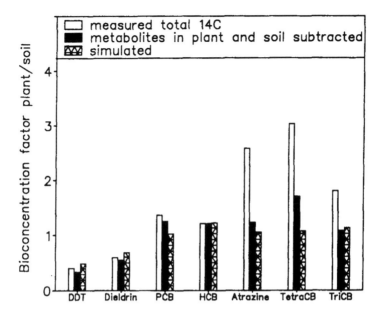

Figure 6. Comparison of experimental and model results for a dynamic simulation.

fugacity is small and far from equilibrium to soil, but even low air concentrations lead to high foliage concentrations because the foliage-to-air bioconcentration factor is very high.

V. DISCUSSION

It is clear that results obtained in small laboratory ecosystems may not be transferred uncritically to real field situations. However, it is possible to study relevant processes under defined conditions and to develop mathematical expressions that may be used for simulations. Generally valid results of this study are:

- The plant/soil bioconcentration factors depend on the ratio of K_{ow}:K_{oc} values and on transfer kinetics.
- Chemicals with a medium K_{ow} are translocated within the plant.
- Chemicals with a high K_{ow} enter foliage via air.

Some important questions are not yet solved, e.g.:

- How much do plant species differ?

Another wide field not touched in this chapter is the metabolism in plants that seems to be much faster than in soil (cf. Figure 1).

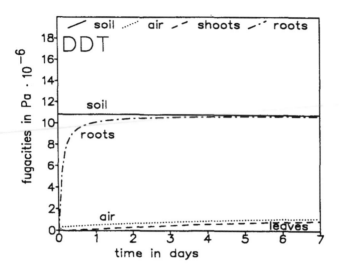

Figure 7. Fugacities of DDT in the model system.

Concentrations of environmental chemicals in plants are much higher than concentrations in water and resemble those in soil and sediment. The hazards caused by toxic residues in the environment can not be estimated when plants are not considered.

ACKNOWLEDGMENT

Thanks are extended to I. Scheunert, E. M. Topp, M. Matthies, R. Schroll, P. Schneider, H. Behrendt, and M. Mansour.

REFERENCES

1. ARGE Elbe. Chlorierte Kohlenwasserstoffe — Daten der Elbe 1980 bis 1982. Wassergütestelle Elbe, Focksweg 32 A, 2103 Hamburg, Germany.
2. Reischl, A., M. Reissinger, and O. Hutzinger. "Organische Luftschadstoffe und ihre Bedeutung für die terrestrische Vegetation," *UWSF-Z. Umweltchem. Ökotox.* 2:32–41 (1989).
3. Topp, E.M. "Aufnahme von Umweltchemikalien in die Pflanze in Abhängigkeit von physikalisch-chemischen Stoffeigenschaften," Doctoral Thesis, Technical University of Munich, Germany (1986).
4. Trapp, S., M. Matthies, I. Scheunert, and E.M. Topp. "Modeling the Bioconcentration of Organic Chemicals in Plants," *Environ. Sci. Technol.* 24 1246–1252 (1990).

5. Riederer, M. "Estimating Transport of Organic Chemicals in the Foliage/Atmosphere System: Discussion of a Fugacity Based Model," *Environ. Sci. Technol.* 24:829–837 (1990).

6. Paterson, S. and D. Mackay. "Modeling the Uptake and Distribution of Organic Chemicals in Plants," in *Intermedia Pollutant Transport: Modeling and Field Measurements,* D.T. Allen and I.R. Kaplan, Eds. (New York: Plenum Publishing Corp., 1989).

7. Bacci, E., D. Calamari, C. Gaggi, and M. Vighi. "Bioconcentration of Organic Vapors in Plant Leaves: Experimental Measurements and Correlation," *Environ. Sci. Technol.* 24:885–889 (1990).

8. Schwarzenbach, R.P. and J. Westall. "Transport of Nonpolar Organic Compounds from Surface Water to Groundwater. Laboratory Sorption Studies," *Environ. Sci. Technol.* 15:1360–167 (1988).

9. Fusseder, A. *Z. Pflanzenernaehr. Bodenkd.* 148:312–334 (1985).

10. Briggs, G.G., R.H. Bromilow, and A.A. Evans. "Relationships Between Lipophilicity and Root Uptake and Translocation of Non-Ionised Chemicals by Barley," *Pestic. Sci.* 13:495–504 (1982).

11. Thompson, N. "Diffusion and Uptake of Chemical Vapour Volatilizing from a Spray Target Area," *Pestic. Sci.* 14:33–39 (1983).

CHAPTER 21

Influence of Soil-Water Ratio on Adsorption-Desorption Kinetics of Isoxaben in Soil

P. Jamet and Dominique Roche

ABSTRACT

The influence of soil-water ratio (Rs/w) on adsorption-desorption kinetics was shown in a quantitative study with isoxaben, a benzamide herbicide, on a silt loam soil (Versailles soil). Such a study made it possible to describe how equilibrium was reached when Rs/w ranged between 0.05 and 1.60, and underlined the interest of the three mathematical models used (hyperbolic model, and two- and three-compartment models). For each model the parameters (especially k21 and k12, the adsorption and desorption rate constants, respectively) were estimated by nonlinear regression. The most suitable model was obtained when the soil was considered as two compartments intended as the fast adsorption phase and the slow adsorption phase, respectively. The increase in the maximum amount of adsorbed isoxaben (Q_{max}) was proportionately lower as Rs/w increased. The equilibrium constant ($K_e = k21/k12$) indicated that isoxaben tended to desorb as Rs/w decreased (Rs/w lower than 0.4). The interest of each mathematical model is discussed.

I. INTRODUCTION

A quantitative study of pesticide adsorption by soil or soil components can be readily and easily conducted using the slurry method. Many attempts have been made to better understand adsorption by using adsorption isotherms, spectroscopic techniques, or applying adsorption theory.

However, the importance of adsorption-desorption kinetics studies, which was first underestimated, is increasingly recognized to gain a more complete knowledge of the dynamics of pesticide in soil. The influence of the soil-water ratio (Rs/w) on isoxaben adsorption and desorption equilibrium was shown in a previous study:[1] estimated Freundlich and Temkin coefficients indicated that isoxaben adsorption decreased as soil-water ratio increased. Then, to complete these methodological aspects, it seems suitable to study also the influence of soil-water ratio on the kinetics of adsorption-desorption; this study was obviously conducted with the same compound on the same soil.

II. MATERIALS AND METHODS

The active ingredient is isoxaben, known as EL-107; it is a benzamide herbicide developed by Eli-Lilly and especially used in Europe for controlling broad-leaved weeds in winter cereals. Isoxaben was used as a model compound in previous studies which indicated that this herbicide, slightly mobile in soil,[2] is not easily desorbed.[3] All the experiments were conducted with Versailles soil, namely, a silt loam. The soil was air-dried, passed through a 2-mm-meshed sieve, and characterized by its organic matter content (1.94%), cation exchange capacity (10.0 meq/100 g), and pH (6.4).

The kinetics study was conducted according to the soil-water ratio chosen (0.05, 0.1, 0.2, 0.4, 0.8, and 1.6); 300 mL of an aqueous solution of [^{14}C]isoxaben was added to a 15, 30, 60, 120, 240, or 480 g sample of Versailles soil and was shaken at room temperature (20°C). At an appropriate time (5, 15, 30, and 45 min; and 1, 2, 3, 5, 7, 10, 24, 48, and 96 hr) two samples were taken, centrifuged at 10,000 gravity (g) for 5 min, and then filtered on a glass fiber filter. For each duplicate, three samples of 0.5 mL each were taken for liquid scintillation counting. The amount of isoxaben adsorbed at time T was calculated using the following equation:

$$x/m = (C_i - C(t)) \cdot V/m$$

where x/m = the amount of isoxaben adsorbed (μg/g)
$\quad\quad Ci$ = initial concentration of isoxaben in solution (μg/mL)
$\quad\quad C(t)$ = concentration of isoxaben in solution (μg/mL) at time T
$\quad\quad V$ = volume of solution (mL)
$\quad\quad m$ = quantity of soil (g)

$$Q \, ads^{(t)} = \frac{Qmax \cdot t}{k + t}$$

Figure 1. Hyperbolic model. (From Jamet et al.[6] ©INRA, Paris, 1985. With permission.)

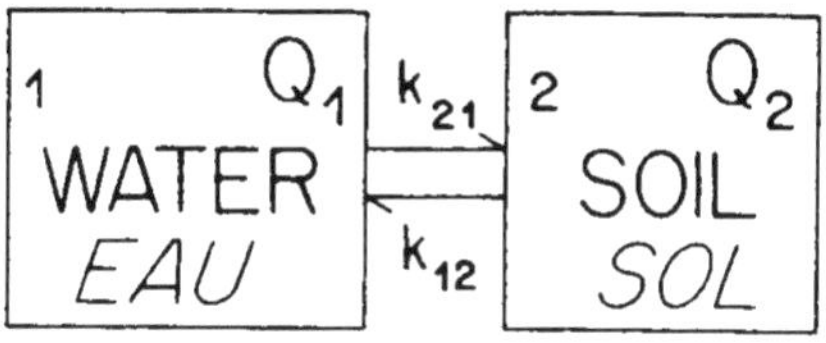

$$\frac{d Q_1}{dt} = -k_{21} Q_1 + k_{12} Q_2$$

$$\frac{d Q_2}{dt} = +k_{21} Q_1 - k_{12} Q_2$$

$$Q_2(t) = \frac{k_{21} Q_0}{k_{21} + k_{12}} \left(1 - e^{-(k_{21} + k_{12})t} \right)$$

Figure 2. Two-compartment model. (From Jamet et al.[6] ©INRA, Paris, 1985. With permission.)

Three mathematical models were used to describe the kinetics of isoxaben adsorption-desorption:

- Hyperbolic model — it is an empirical and descriptive model, described by Biggar et al.[4] (Figure 1).
- Two- and three-compartment models — these are explanatory models proposed by Leistra and Dekkers[5] (Figures 2 and 3).

The two-compartment model involved a first-order rate equation for adsorption-desorption; the three-compartment model involved simultaneous rapid and slow adsorption phases. These three models have been used previously and compared to the study the kinetics of adsorption-desorption of tioclorine by Jamet et al.[6] or carbofuran by Achick et al.[7] Nonlinear regression was used to estimate the numerical parameters of each model.[6,7] Then, adsorption rate curves of each model were drawn with a Schlumberger plotter using GPGS-F software running on a DPS-8/MULTICS computer.

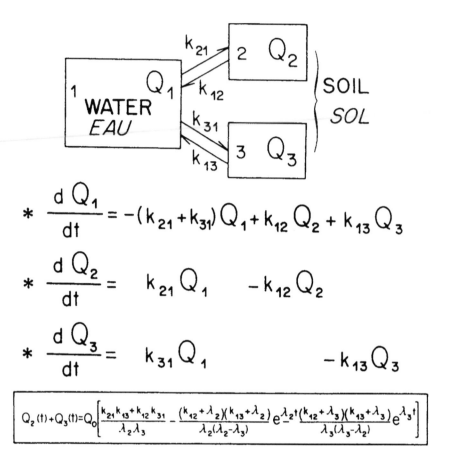

$$* \quad \frac{d Q_1}{dt} = -(k_{21} + k_{31}) Q_1 + k_{12} Q_2 + k_{13} Q_3$$

$$* \quad \frac{d Q_2}{dt} = \quad k_{21} Q_1 \quad - k_{12} Q_2$$

$$* \quad \frac{d Q_3}{dt} = \quad k_{31} Q_1 \qquad\qquad - k_{13} Q_3$$

$$Q_2(t) + Q_3(t) = Q_0 \left[\frac{k_{21} k_{13} + k_{12} k_{31}}{\lambda_2 \lambda_3} - \frac{(k_{12} + \lambda_2)(k_{13} + \lambda_2)}{\lambda_2 (\lambda_2 - \lambda_3)} e^{\lambda_2 t} + \frac{(k_{12} + \lambda_3)(k_{13} + \lambda_3)}{\lambda_3 (\lambda_3 - \lambda_2)} e^{\lambda_3 t} \right]$$

Figure 3. Three-compartment model. (From Jamet et al.[6] ©INRA, Paris, 1985. With permission.)

III. RESULTS AND DISCUSSION

Only the main results of this study can be given and discussed here. Estimations of Q_{max} (maximum amount of isoxaben adsorbed) and estimated values for the parameters of the two- and three-compartment models are given in Tables 1 and 2, respectively. Some graphs of reaction rates are shown in Figures 4, 6, and 7.

The interest of the models is demonstrated in Figure 4. As observed in previous studies,[6,7] the three-compartment model provided the best fit of adsorption-desorption of isoxaben kinetics as a function of time. The coefficients of determination (see some examples in Figure 4) were always higher than those obtained for the other two models. The two-compartment model did not provide a good description of the adsorption-desorption phenomena; this model reaches

Table 1. Adsorption Kinetics of Isoxaben by Versailles Silt Loam and Estimations of the Maximum Amount of Isoxaben Adsorbed for Various Soil-Water Ratios

Soil-water ratio (Rs/w)	Q_o (μg)	Maximum amount of adsorbed isoxaben (Qmax, μg)		
		Hyperbolic	2 Compartment model	3 Compartment model
0.05	69.52	16.26	15.46	19.79
0.10	72.30	23.77	22.94	30.04
0.20	68.87	30.31	28.92	34.82
0.40	69.34	41.48	39.87	43.10
0.80	64.99	48.02	46.72	51.01
1.60	65.52	54.63	51.82	57.40

equilibrium too rapidly and then is only reliable for describing the beginning of the adsorption, i.e., the fast adsorption phase. The hyperbolic model can allow a satisfactory description of the adsorption-desorption kinetics. Because of the possibility of using linear regression,[4] such an elementary model provides a good estimation of Q_{max} (see below). However, only the two- and three-compartment models allowed the estimation of adsorption and desorption reaction rate constants.

Because of the soil structure (soil porosity) and soil components (organic matter and clays), adsorption equilibrium was not immediately reached (as can be seen in Figure 4). In all the cases the three-compartment model gave the highest values for Q_{max} (Table 1); the estimated values from the two-compartment model were always lower than those from the three-compartment model. The hyperbolic model provided intermediate values for Q_{max}.

The adsorption-desorption kinetics of isoxaben involved variation of the estimated values for the adsorption and desorption reaction rate constants demonstrating the influence of the Rs/w. Generally, the adsorption reaction rate constant (k21) increased as Rs/w increased, whereas the desorption reaction rate constant (k21) decreased. Thus, the equilibrium constant Ke (Ke = k21/k12) for the two-compartment model ranged from 0.29 (Rs/w = 0.05) to 3.78 (Rs/w = 1.60). To illustrate these variations, the equilibrium constant Ke was plotted against Rs/w (Figure 5). It is apparent that when the Rs/w ranges between 0.05 and 0.2, the equilibrium constant K_e is lower than 1; thus isoxaben tends to desorb. When the Rs/w exceeds 0.2, K_e becomes higher than 1; then isoxaben tends to adsorb (Table 2, Figure 5). Such a result was confirmed by the three-compartment model, both by the K_{ef} and K_{es}, the equilibrium constants for the fast adsorption and slow adsorption compartments, respectively. Thus, the equilibrium constant K_e, where $K_e = K_{ef} + K_{es}$, for the three-compartment model (Table 2, Figure 6) corroborated this trend. However, the values of K_e obtained for the three-compartment model showed that isoxaben tended to desorb only for the two lowest values of Rs/w, i.e., 0.05 and 0.1. In all the cases, the values of K_e calculated from the three-compartment model were higher than those from the two-compartment model, especially for the highest values of Rs/w.

Table 2. Adsorption Kinetics of Isoxaben

Soil-water ratio (Rs/w)	Two-compartment model[a]			Three-compartment model[a]						
				Fast adsorption			Slow adsorption			
	k21	k12	K_e[b]	k21	k12	K_{ef}[c]	k31	k13	K_{es}[d]	K_e[e]
0.05	1.73	6.05	0.29	3.51	15.25	0.23	0.005	0.033	0.15	0.38
0.10	3.24	6.98	0.47	4.60	11.63	0.40	0.007	0.023	0.30	0.70
0.20	3.07	4.24	0.72	4.73	8.15	0.58	0.019	0.043	0.44	1.02
0.40	4.36	3.22	1.35	6.13	6.40	0.96	0.128	0.187	0.68	1.64
0.80	7.19	2.81	2.56	8.76	4.39	2.00	0.091	0.055	1.66	3.65
1.60	3.66	0.97	3.78	7.36	5.30	1.39	0.455	0.080	5.69	7.08

[a] Parameters of the two- and three-compartment models are estimated for different soil-water ratios.
[b] Equilibrium constant for the two-compartment model K_e = k21/k12.
[c] Equilibrium constant for the fast adsorption compartment K_{ef} = k21/k12.
[d] Equilibrium constant for the slow adsorption compartment K_{es} = k31/k13.
[e] Equilibrium constant for the three-compartment model K_e = K_{ef} + K_{es}.

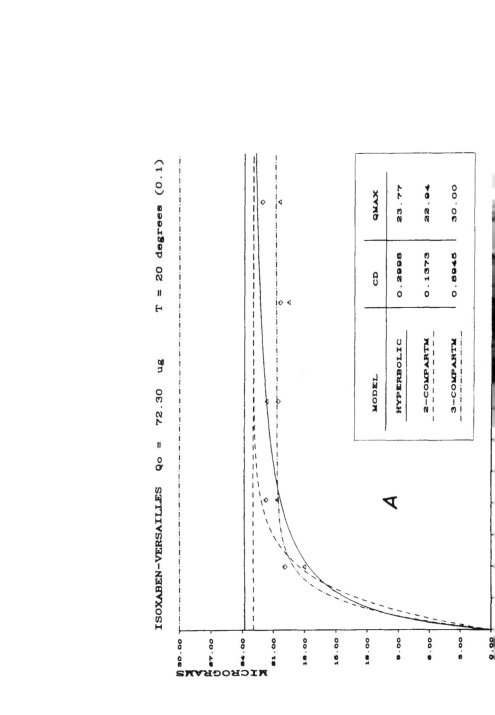

ISOXABEN–VERSAILLES   Qo = 72.30 ug   T = 20 degrees (0.1)

MODEL	CD	QMAX
HYPERBOLIC	0.2998	23.77
2-COMPARTM	0.1373	22.94
3-COMPARTM	0.8945	30.00

A

MICROGRAMS

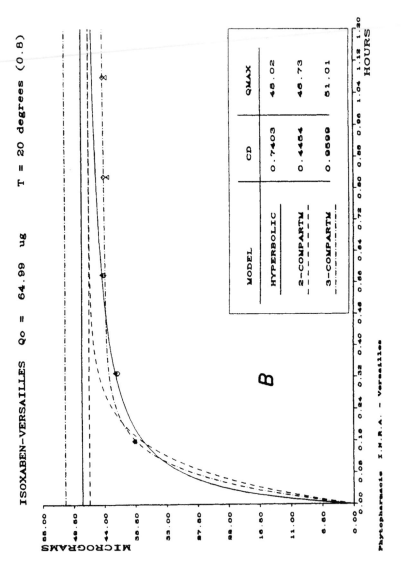

Figure 4 (continued).

ISOXABEN – Ke and Rs/w

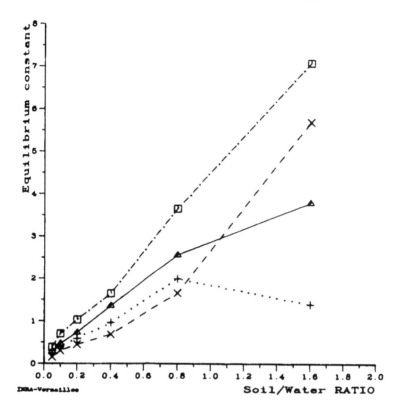

Figure 5. Adsorption-desorption kinetics of isoxaben on Versailles silt loam soil. Equilibrium constants for two- and three-compartment models for soil-water ratios from 0.05 to 1.60 —— K_e for the two-compartment model; — · — K_e for the three-compartment model; · · · · K_{ef} for the fast adsorption compartment; and — — — K_{es} for the slow adsorption compartment.

The resulting apparent role played by slow adsorption is shown in Figure 6. For the lowest values of Rs/w (i.e., 0.05 and 0.1), the variation of the total amount of isoxaben adsorbed can be described in the first hour by the fast adsorption compartment (Figure 6). As Rs/w increased, the simulation indicated that the amount of isoxaben adsorbed by the slow adsorption compartment increased as Rs/w increased (Figure 6). Figure 7 shows that a rapid adsorption quickly removes isoxaben from the aqueous solution (fast adsorption phase) and then releases it, while isoxaben is removed by an adsorption phenomena which proceeds more slowly (slow adsorption phase). Since the amount of isoxaben adsorbed by the slow adsorption compartment increased as Rs/w increased, it can be assumed that this compartment might represent the internal diffusion necessary to reach adsorbent sites in soil microporosity. At the lowest Rs/w, soil

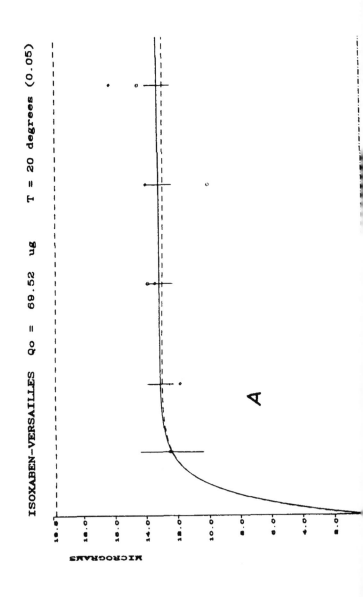

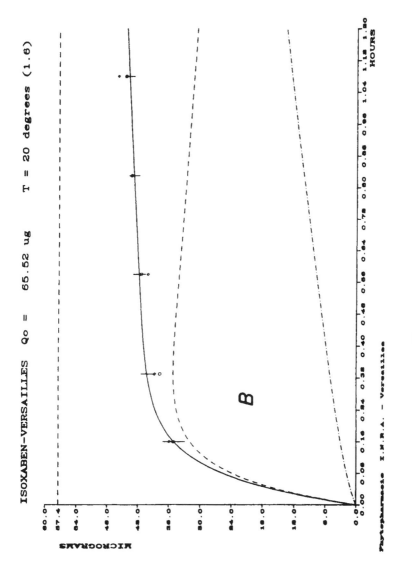

Figure 6 (continued).

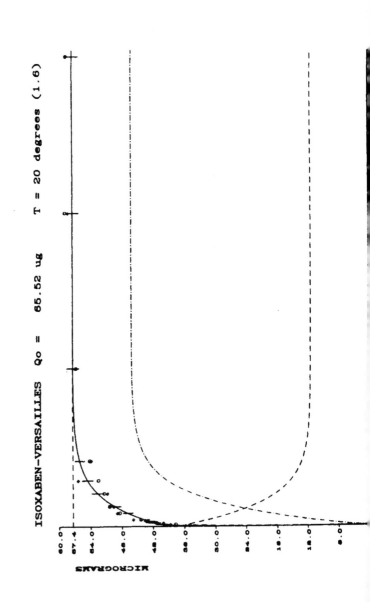

aggregates were quickly destroyed; then adsorbent sites were more easily accessible.

IV. CONCLUSION

The adsorption of pesticides by such a heterogeneous material as soil is a natural phenomenon among the most difficult to interpret. Observations on soil systems are nearly always obtained on the macroscopic level. However, mathematical models may be powerful tools for describing and even for explaining adsorption-desorption kinetics of pesticides in soils. The two-compartment model can be used in a satisfactory manner for the fast adsorption phase, assuming a first-order reaction rate for low values of the Rs/w as for pesticide-water-sediment interactions. For higher values of Rs/w, close to field conditions, the three-compartment model is needed for a better description and interpretation of adsorption-desorption phenomena.

ACKNOWLEDGMENTS

The authors thank Guy Decoux (Laboratorie de Biométrie, INRA Versailles) for his help in writing the FORTRAN programs to pilot the plotter, and DOW ELANCO company for kindly suppling [^{14}C]isoxaben.

REFERENCES

1. Jamet, P. and D. Hoyoux-Roche. "Influence du rapport sol/eau lors de l'etude quantitative de l'adsorption et de la désorption des pesticides," *Symposium Pesticides-Soils: Methodological Aspects of the Study of Pesticide Behaviour in Soil,* P. Jamet, Ed. (Versailles: Service des publications INRA, 1989).
2. Jamet, P. and J.C. Thoisy-Dur. "Pesticide Mobility in Soils: Assessment of the Movement of Isoxaben by Soil Thin-Layer Chromatography," *Bull. Environ. Contam. Toxicol.* 41:135–142 (1988).
3. Jamet, P., A. Copin, R. Deleu, J.C. Thoisy-Dur, D. Hoyoux, and G. Carletti. "Aspects méthodologiques de l'étude de l'adsorption et de la désorption des herbicides dans le sol," *EWRS Symposium: Factors Affecting Herbicidal Activity and Selectivity,* (Wageningen: Pays-Bas, 1988), 263–268.
4. Biggar, J., U. Mingelgrin, and M.W. Cheung. "Equilibrium and Kinetics of Adsorption of Picloram and Parathion with Soils," *J. Agric. Food Chem.* 26:1306–1312 (1978).
5. Leistra, M. and W.A. Dekkers. "Some Models for the Adsorption Kinetics of Pesticides in Soil," *J. Environ. Sci. Health* 2:85–103 (1977).

6. Jamet, P., J.-C. Thoisy, and C. Laredo. "Etude et modelisation de la cinétique d'adsorption des pesticides dans le sol" *Symposium INRA-FAO Comportement et effets secondaires des pesticides dans le sol* (Versailles: Service des publications INRA, 1985), 135–146.

7. Achik, J., M. Schiavon, and P. Jamet. "Study of Carbofuran Movement in Soils. II. Describing the Kinetics of Carbofuran Adsorption with Different Mathematical Models," 6th *Conference of Chemistry for Protection of Environment*, Turin: 15–18 Sep 1987. *Environ. Sci.* 17:81–88 (Elmsford, NY: Pergamon Press, 1991).

CHAPTER 22

Toxicity and Metabolism of Cypermethrin in Earthworms

R. Viswanathan and Zhen-Hui Zhou

I. INTRODUCTION

The cyanopyrethroid insecticide, cypermethrin, is widely used in the control of a variety of insect pests. It acts as nerve poison in the target organisms.[1] As it is highly lipophilic and observed to build up quickly bound residues in soil, which degraded only slowly to carbon dioxide,[2] in recent years investigations have been conducted in a number of laboratories to determine its potential for accumulation, toxicity, and biotransformation in earthworms as an important group of soil nontarget organisms.

II. EXPERIMENT

The toxicity of cypermethrin has been investigated according to an artificial soil test procedure using *Eisenia foetida*.[3] In the tests, the technical grade substances, Ripcord and Fastac were employed.

The uptake and biotransformation of the substance by *E. foetida* were determined in our laboratory using a closed model ecosystem set up with a 30-cm desiccator in phytotron.[4] The experiments were run for 2 weeks. The desiccator contained 11 one-week-old maize seedlings and 15 *E. foetida* species (average

0-87371-616-7/93/$0.00 + $.50

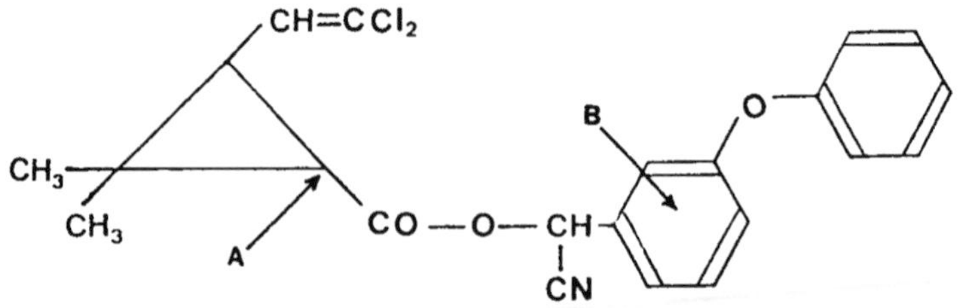

Figure 1. Cypermethrin ((RS)-α-cyano-3-phenoxybenzyl (1RS)-cis, trans-3-(2,2-dichloro-vinyl)-2,2,-dimethyl cyclopropanecarboxylate). ^{14}C-Labelling (A) in acid moiety and (B) in alcohol moiety.

weight 460 mg) in a sandy loam soil that was enriched with 5% cattle manure and that had been treated with 5.5-ppm purified ([^{14}C] alcohol) cypermethrin (Figure 1). Four 1-week-old maize seedlings in an untreated soil in a dish also were maintained simultaneously in the same desiccator for control purposes. In the course of the experiments, fresh filtered air (Rate about 16.5 mL/min) was passed through the system with the help of a membrane pump. To determine the volatilized substance and the mineralization product (^{14}CO$_2$), respectively, traps with a mixture of ethyleneglycol monomethylether and Permablend and that with a mixture of Carbosorb II and Permafluor were attached to the desiccator in the order serially.

The potential for accumulation and biotransformation were studied also in the earthworm species, *Lumbricus terrestris* and *Allolobophora caliginosa*.[5] The experiments were conducted initially by exposing them in a loamy coarse sand soil containing 1-ppm ([^{14}C]alcohol) cypermethrin for a period of 8 weeks and then transferring them to untreated soil. In a subsequent experiment, *A. caliginosa* was only exposed to 10-ppm soil concentrations of ([^{14}C]alcohol)- as well as ([^{14}C]acid) cypermethrin for 8 weeks.

III. RESULTS AND DISCUSSION

The toxicity data of cypermethrin determined with *E. foetida* are shown in Table 1. It is clear from the data that neither compound caused significant mortality at substrate concentrations up to 100 ppm.

Inglesfield[6] citing literature has reported that cypermethrin applied in the field plots at the prescribed rates would pose no hazard to either the abundance or the biomass of the worms, *Lumbricus* spp. and *Allolobophora* spp., and only occasional reduction in numbers and biomass were observed at very high rates of applications.

The *E. foetida* that were exposed to substrate containing 5.5-ppm ([^{14}C]alcohol) cypermethrin accumulated about 2.2% of the total applied radioactivity in a 2-

Table 1. 'Artificial Soil Test' Using *Eisenia foetida*

Treatment	Dose (mg/kg)	Mean weight of worms (mg)	Mortality (%)
Cypermethrin	0.1	330	5.0
(Ripcord)	1.0	330	7.5
	10	340	5.0
	100	330	5.0
WL85871	0.1	360	2.5
(Fastac)	1.0	360	5.0
	10	340	2.5
	100	350	5.0

Source: Inglesfield, C. *Bull. Environ. Contain. Toxicol.* 33:568–570 (1984). With permission.

week duration. In terms of bioconcentration, this amounted to about 2.6 times the ambient concentration.

Most of the accumulated residues were readily organic soluble, and more than 90% of the organic soluble had been biotransformed. Thin-layer chromatography (TLC) analysis of the extract revealed the presence of at least 7–8 metabolites (Figure 2 and Table 2). Two of the components with Rf values 0.55 and 0.65, respectively, corresponded to 3-phenoxybenzaldehyde and 4'-methoxy-3-phenoxybenzaldehyde.

The *L. terrestris* and *A. caliginosa* that were maintained in soil treated with 1-ppm ([^{14}C]alcohol) cypermethrin had accumulated about 8-fold and 30-fold, respectively, one of the radioactive residues in 8 weeks. These residues were not eliminated following transfer of the worms to untreated soil for 8 weeks (Figure 3). When *A. caliginosa* were maintained for 8 weeks in soil treated with 10-ppm ([^{14}C]alcohol) cypermethrin and that with 10-ppm ([^{14}C]acid) cypermethrin, respectively, the acid worms exhibited much lesser accumulation of the substance than the alcohol worms.

In both the acid and alcohol worms only less than 1% of the accumulated residues was hexane extractable. Aqueous methanol (1:1 by volume), however, solubilized 48 and 75%, respectively, of the residues from the acid and alcohol worms. The accumulated residues were observed to consist of a complex mixture of conjugates of 3-phenoxybenzoic acid and (1RS) cis, trans-3-(2,2-dichlorovinyl)-2,2-dimethylcyclopropane carboxylic acid. A major constituent of the mixture of conjugates in the alcohol worms was identified as N^1, N^{12}-di-(3-phenoxybenzoyl) spermine.

IV. CONCLUSION

Cypermethrin is accumulated by earthworms to different extents from the treated soils. The level of bioconcentration, however, appears to be dependent both on the species and the substrate concentration. Laboratory and field observations indicate that this insecticide may not pose a hazard to either the

Figure 2. Autoradiogram of TLC separation of ([¹⁴C]alcohol) cypermethrin metabolites from benzene extracts of exposed *Eisenia foetida.*

Table 2. The Over-All Degradation/Conversion (in Percentage) of [14C]Cypermethrin in the Closed Ecosystem in Relation to Total Application

Samples	Unconverted [14C]cypermethrin	TLC of Degradation/Conversion Products (Rf values)[a]						Bound and conjugated residues
		0.00–0.10	0.17	0.30	055	0.65	0.83	
Soil	74.1	0.63	0.06	0.23	3.8	0.33	—	15.5
Earthworm	0.18	0.51	0.25	0.27	0.18	0.12	0.12	0.55
Maize	0.16				0.05			1.2
Air 14CO2					1.7			

Source: Viswanathan.[4]

[a] Rf values after successive runs in benzene + n-hexane (3 + 2) 3x and benzene + ethylacetate + methanol (15 + 5 + 1) 1x.

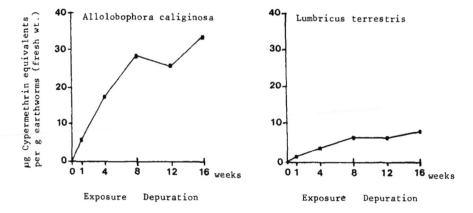

Figure 3. The radioactive residues (expressed as cypermethrin equivalents) in earthworms maintained in soil containing ([^{14}C]alcohol) cypermethrin at an initial concentration of 1 ppm (From Curl, E.A., P.J. Edwards, C. Elliott, and J.P. Leahey. *Pestic. Sci.* 20:207–222 [1987]. With permission.)

population or the biomass of earthworms when used at the recommended rates. Earthworms biotransform cypermethrin to more than 90% within a short period of time. The metabolites result from two fragments produced by cleavage of the ester linkage. The nature of the metabolites seems to indicate that the lipophilicity of the compound is not the real cause of its bioaccumulation.

REFERENCES

1. Miyamoto, J. *Chirality and Biological Activity,* (New York: Alan R. Liss, 1990), pp. 153–168.
2. Demoute, J.-P. *Pestic. Sci.* 27:375–385 (1989).
3. Inglesfield, C. *Bull. Environ. Contam. Toxicol.* 33:568–570 (1984).
4. Viswanathan, R. In: *Ecotoxicology of Earthworms,* P.W. Greig-Smith, H. Becker, P.J. Edwards, and F. Heinbach, Eds. (Andover, Hants, England: Intercept, 1992) pp. 217–219.
5. Curl, E.A., P.J. Edwards, C. Elliott, and J.P. Leahey. *Pestic. Sci.* 20:207–222 (1987).
6. Inglesfield, C. *Pestic. Sci.* 27:387–428 (1989).

CHAPTER 23

Determination of the Rate Constant k_{OH} (Air) Using Freon 113 as an Inert Solvent

W. Klöpffer and E.-G. Kohl

I. INTRODUCTION

Many pesticides partially evaporate during and after application and therefore enter into the lower troposphere. Their fate in this compartment of the environment depends on physical distribution and chemical degradation processes. The most important distribution process which influences the photochemical degradation (at least in a dry atmosphere) occurs between the gas phase and the particle phase and can be approximately described by Junge's formula (Equation 1):[1,2]

$$\phi = \frac{K \cdot \Theta}{p_{20} + K \cdot \Theta} \tag{1}$$

where ϕ = adsorbed fraction of the chemical
p_{20} = vapor pressure at 20°C (Pa)
Θ = particle surface/volume of air (cm^2/cm^3)
K = empirical constant ≈ 18 Pa cm

Equation 1 clearly shows that substances with vapor pressures down to about 10^{-7} Pa may occur in the vapor phase in an atmosphere of low total particle surface. On the other hand, vapor pressures above 10^{-3} Pa at room temperature indicate molecular distribution even in highly polluted air masses.

At lower temperatures, adsorption is favored due to the exponential decrease of vapor pressure. Therefore, the upper vapor pressure limit of the range of

0-87371-616-7/93/$0.00 + $.50

chemicals called[3] semivolatile organic compounds (SOC) should be taken at about 1 Pa. The SOC may occur either in the gas phase or in the adsorbed state, depending on temperature, particle surface, and vapor pressure. The total range, therefore, extends over seven orders of magnitude in pressure, from about 10^{-7} to 1 Pa. Most pesticides and many other chemicals belong to this range and thus to the class of SOC.

In order to estimate the degradation behavior of these compounds in the atmosphere, we have to know the rate constants of direct and indirect photochemical reactions acting as sinks in the gas phase and in the adsorbed state.[4,5] In the following, we shall only discuss reactions in the gas phase, which in the case of SOC cannot be studied directly for technical reasons.[5] Only in very favorable cases, a lower limit of about 10^{-2} Pa can be reached, using the smog chamber method.[6]

We wish to report on a method of determination of k_{OH}, the biomolecular rate constant of the most universal reaction (Equation 2) which initiates the degradation of most chemicals in the atmosphere:

$$M + OH \rightarrow Products \tag{2}$$

$$-d[M]/dt = k_{OH} [M] [OH] \tag{3}$$

The method consists of measuring relative OH-reaction rates in the inert solvent Freon 113 (1,1,2-trichloro-1,2,2-trifluoroethane). This technique was proposed by us in 1980[7] and was shown to be practicable in preliminary experiments.[8] Meanwhile, Dilling and co-workers[9] have reported about 30 relative OH-reaction constants obtained with an improved version of this method, which nicely fit into the relative order of k_{OH} values measured in the gas phase. The relative k_{OH}-data measured in the inert solvent can therefore easily be converted into absolute gas-phase data. The results reported here essentially confirm the results obtained by Dilling et al.[9] as well as our earlier predictions of the usefulness of this indirect method of determination of k_{OH} (gas).[7,8]

II. PRINCIPLE OF THE METHOD

A. Generation of OH Radicals

In order to measure relative OH-reaction rates in Freon 113, a stationary concentration of OH radicals is produced by UV photolysis of hydrogen peroxide (H_2O_2) at wavelengths ≥ 300 nm (see Secion III and Figure 1). The photolysis of H_2O_2 mainly has been studied in the gas phase[10-12] and in water.[13-15] We assume that the quantum efficiency $\phi(-H_2O_2)$ in freon 113 is nearer to that in water ($\phi = 0.5$) than to the one observed in the gas phase ($\phi = 1.0$) due to the cage effect. On the other hand, no dissociation of OOH into $O_2^-\cdot$ and H^+

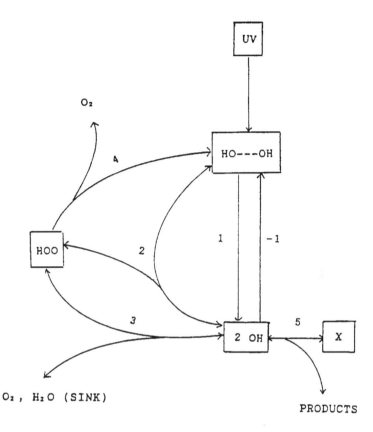

Figure 1. Reaction scheme of the photolysis of hydrogenperoxide in air and in nonpolar solvents.

should be possible in the nonpolar solvent; as in the gas phase, we only have to consider OOH as the main secondary species during the photolysis of H_2O_2.

In Figure 1 the OH radicals generated in reaction (1) may react with the parent compound H_2O_2 to give OOH (2) with the secondary product OOH (3) or with a foreign substance X, which may be an impurity, the test-, and the reference compound (5). The recombination (-1) is not efficient due to the very small concentration of the highly reactive OH radicals. The OOH radicals partially recombine to give H_2O_2 (4). Using ordinary UV intensities it is possible to maintain a sufficiently high OH concentration in order keep the reaction times in the order of a few hours to days.

B. Relative Reaction Rates

The principle of measurement consists of measuring the decay of the test substance and at least one reference compound dissolved in Freon 113 which is saturated with H_2O_2.

At least one main run and two different blanks have to be performed:

Main run:	Test substance(s) + reference substance + H_2O_2 + UV
Blank 1:	Test substance(s) + H_2O_2 (keeping the reaction mixture in the dark)
Blank 2:	Test substance(s) + UV (irradiation of the solution of the test substance in freon 113 without H_2O_2 added)

Blank 1 assures the absence of any dark reaction or the application of a correction. Blank 2 allows quantification of any direct photochemical reactions of the test substance. Rapid photolysis of the test substance, as indicated by fast disappearance during blank 2, prevents the determination of k_{OH} by this method. Such compounds are also likely to be photolyzed in the atmosphere very rapidly, so that the OH reaction should play a minor role as a sink in these cases.

The decay of test and reference substances is, in general, exponential (if [OH] $\approx$ constant). In this case, the relative rate constant of the test substance is simply given by the ratio of slopes in the semilogarithmic plot. The decay curve of the test substance has to be corrected according to the blank tests (the reference substance should not react in the blank tests; in case it does, a correction has to be applied). If the decay is not nonexponential, the evaluation after Cox[16] can be applied (Figure 2).

III. EXPERIMENT

The apparatus used is similar to the one described by Dilling et al.[9] It is a slightly modified version of an apparatus produced by NORMAG (Hofheim/Taunus, Germany; see Figure 3) which has been described in a recent textbook on photochemistry.[17] In this apparatus, the solution to be irradiated is circulated by means of a pump. The irradiation is performed with a water-cooled, 125 W, Hg high pressure lamp (TQ 150, produced by Heraeus, Hanau, Germany) equipped with a Pyrex filter. The temperature was kept constant during the experiments at 25°C. During irradiation, the main vessel is surrounded with a metal cylinder in order to protect the operator from UV radiation.

The main function of the side arm (Figure 3) is continuous recharging of the solution with H_2O_2. For this purpose, a layer of 5 mL H_2O_2 (ca. 90%, Degussa) is floating on the circulating solution on top of the side arm. It is permanently in contact with the moving solution and thus supplies H_2O_2 for the photolytic process in the main vessel. The charging is necessary[9] since H_2O_2 is only sparingly soluble in freon 113 (50 mg/L at 25°C). The UV absorption above 300 nm is weak ($\epsilon_{300} \approx 1$ L/mol/cm) but sufficient in order to allow the photolysis by Hg lines at 302, 313, 334, and 366.

The handling of 90% H_2O_2 requires security measures which are described in References 18–20. The solvent freon 113 is an ozone-depleting substance which has to be disposed as hazardous waste.

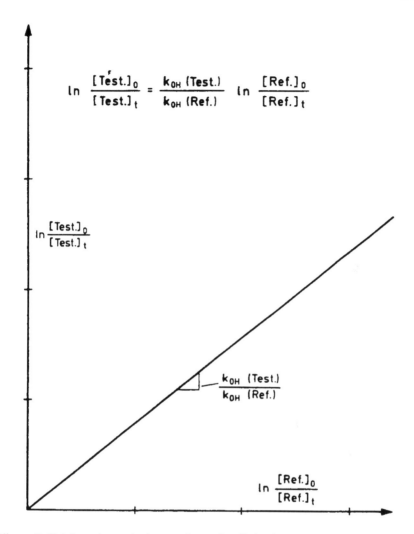

Figure 2. Relative rate constants according to Cox;[16] also for nonexponential decay.

IV. RESULTS

A. Comparison with Published Data

In Figure 4, the results of an experiment with cyclohexane, the preferred reference substance by Dilling et al.,[9] and trichloroethene ([Tri], upper curve) are shown. The decay of cyclohexane had to be corrected for decay in the blanks. Cyclohexane, therefore, is not an ideal reference substance (later on we preferred toluene). The decay of Tri needed no correction.

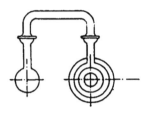

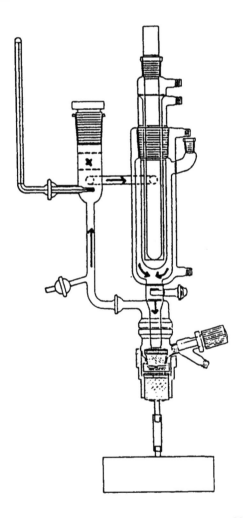

Figure 3. Irradiation apparatus (NORMAG). x: 90% hydrogen peroxide.

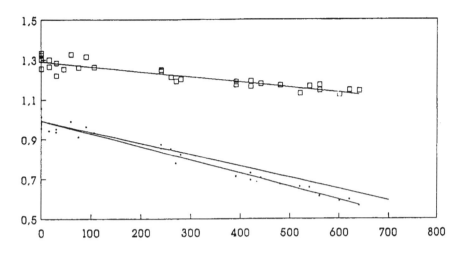

Figure 4. OH reaction in Freon 113. Trichloroethene (upper curve) and cyclohexane (lower curve); log concentration (mg/L) vs irradiation time (min).

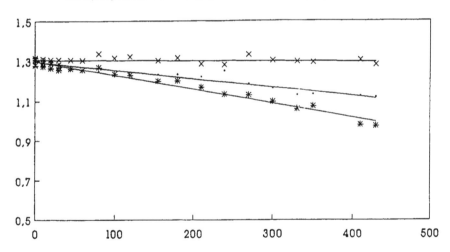

Figure 5. OH reaction in Freon 113. 1,2,4-Trichlorobenzene (upper curve), cyclohexane (middle curve), and p-cresol (lower curve); log concentration (mg/L) vs irradiation time (min).

In Figure 5, the reactions of p-cresol, 1,2,4-trichlorobenzene, and cyclohexane (reference) are compared. As can be seen, the decay of p-cresol can be evaluated, whereas for trichlorobenzene the irradiation time was too short.

In Table 1, the data obtained for those compounds, for which k_{OH} data exist, are compiled and compared with literature.

As can be seen the k_{OH} value of p-cresol measured in Freon 113 is evidently too low. For the slowly reacting substances trichlorobenzene and methylchlo-

Table 1. Results Obtained with Substances of Known k_{OH} (Gas)

Substance	k_{rel}[a]	This work (cm³/sec)	Dilling et al[9] (cm³/sec)	Measured[21b] (cm³/sec)
p-Cresol	1	7.4×10^{-12}		$(4.4 \pm 0.5) \times 10^{-11}$
1,2,4-Trichlorobenzene	0	$<10^{-12}$	7.8×10^{-13}	$(5.32 \pm 0.5) \times 10^{-13}$
EtOAc	0.21	1.5×10^{-12}	2.1×10^{-12}	$(1.82 \pm 0.36) \times 10^{-12}$
1,1,1-Trichloroethane	0	$<10^{-12}$	$\approx 10^{-14}$	$(1.19 \pm 0.20) \times 10^{-14}$
Trichloroethene	0.46	3.4×10^{-12}	7.7×10^{-13}	$(2.36 \pm 0.7) \times 10^{-12}$

[a] Relative to cyclohexane.
[b] Directly in the gas phase.

roform, the irradiation times used were too short; the data obtained by Dilling et al.,[9] however, compare well with gas-phase data.[21]

Ethylacetate (EtOAc) shows a good coincidence with the measured gas-phase data both in our experiment and in Dilling's.[9] In the case of Tri, the value reported here is much closer to the gas-phase k_{OH}[21] than the one reported by Dilling.[9] This discrepancy is due to a high blank 2 correction of 80% reported by Dilling,[9] which seems to be an artefact and is absent in our experiment.

B. Results Obtained with Semivolatile Organic Compounds (SOC)

As a "praxis test" three SOC were measured, for which no gas phase k_{OH} data are available. The results are compiled in Table 2.

The herbicide atrazine belongs with a vapor pressure at 20°C:

$$P_{20} = 1.4 \cdot 10^{-4} \ Pa^{22}$$

clearly to the class of SOC. Under approximately rural conditions ($\theta \approx 10^{-6}$ cm²/cm³),[2] the fraction adsorbed at the particles should be $\phi \approx 0.1$ according to Equation 1.

The rate constant k_{OH} has been measured with toluene as the reference substance. According to Equation 4, the chemical (OH−) lifetime $\tau(OH)$ can be calculated, using the global average of $\langle [OH] \rangle = 5 \times 10^5 \ cm^{-3}$:[21]

$$\tau(OH) = \frac{1}{k_{OH} \cdot \langle [OH] \rangle} = \frac{1}{k_{OH}^I} \tag{4}$$

According to this measurement, atrazine reacts easily with OH radicals in the gas phase, the chemical (OH−) lifetime being only 1.7 days. This value is an upper limit to the total chemical lifetime, which is given by the reciprocal sum of all pseudo first-order rate constants k_x^I (Equation 5):

$$\tau = \frac{1}{\Sigma \ k_x^I} \tag{5}$$

$$X = OH, \ h\nu, \ O_3, \ NO_3$$

Table 2. Results Obtained with SOC of Unknown k_{OH} (gas)

Substance	k_{OH} (gas phase) (cm³/sec)	τ(OH) (d)
Atrazine	14×10^{-12}	1.7
DEHP	7.4×10^{-12}	3
OCDD	7.0×10^{-12}	3.3

It should be mentioned, however, that τ may actually be much longer in unfavorable climatic situations, especially at low solar irradiance.[23] The same caveat is, of course, appropriate for the following results obtained with di(2-ethylhexyl-phthalate) (DEHP) and octachlorodibenzo-p-dioxin (OCDD).

The vapor pressure of the common plasticizer DEHP amounts to:

$$p_{20} = 6.0 \cdot 10^{-6} \text{ Pa}$$

The fraction adsorbed according to Equation 1 is calculated for average rural particle surface[2] as $\phi = 0.75$. For DEHP, therefore, degradation in the adsorbed state should be more relevant than for atrazine. In the gas phase, DEHP should degrade with an average chemical (OH$^-$) lifetime of:

$$\tau(\text{OH}) \approx 3 \text{ days}$$

Behnke et al.[24] investigated the photodegradation of DEHP adsorbed on various small particles in a 2.4 m³ aerosol chamber.[25] On SiO$_2$ particles, a decay of $\tau \approx 3.5$ hr was observed. Since the strong aritficial irradiation produced a photosmog with a 10-fold higher OH concentration ([OH] = 5×10^6/cm³) compared to the global average, the chemical lifetime in the adsorbed (ads.) state amounts to:

$$\tau(\text{OH,ads.}) \approx 1.5 \text{ days}$$

OCDD was studied as a less toxic model compound of the more volatile lower chlorinated PCDD (Figure 6).

The vapor pressures at 20 and 25°C (reported by Rippen[22] vary in a wide range between 8.7×10^{-6} Pa and 1.1×10^{-10} Pa. OCDD, therefore, may or may not belong to the class of SOC. Experimental studies show that most likely the lower chlorinated PCDD with five and less Cl atoms are desorbed from the particle phase, whereas OCDD is not.[26] OCDD is degraded by OH as well as by direct photochemical reactions (see Figure 6, photolysis, blank 2). Photolysis has been observed previously at high temperature.[27] The reaction with OH:

$$\tau(\text{OH}) \approx 3 \text{ days}$$

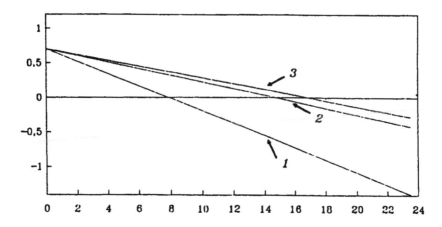

Figure 6. OCDD in Freon 113. Main run (1); blank 2 (2); OH-reaction (main run — blank 2). (3). Log relative concentration vs time (hr).

however, is a surprise considering the fully halogenated structure of OCDD. A recent estimate of τ(OH) by Atkinson[29] amounts to 9.6 days. If the lower halogenated congeners react equally efficiently, this reaction would constitute a sink for the volatile fraction of the PCDD. An estimated (estim.) value for 2,3,7,8-TCDD has been reported by Atkinson:[28]

$$k_{OH}(estim.) \approx 9 \cdot 10^{-12} \text{ cm}^3/\text{s}$$

Much more experimental work will be needed, however, to assess the importance of the gas-phase reaction for SOC and to collect enough data for the rate constants.

ACKNOWLEDGMENT

This work has been performed under contract No. 106 02 57 for Umweltbundesamt, Berlin. We express our thanks to Dr. Petra Greiner and to Mr. H.-J. Poremski for their help and continued interest.

REFERENCES

1. Junge, C.E. "Basic Considerations About Trace Constituents in the Atmosphere as Related to the Fate of Global Pollutants," *Prepr. Div. Environ. Chem. Am. Chem. Soc.* 15(1):4 (1975).
2. Gill, P.S., T.E. Graedel, and C.J. Weschler. "Organic Films on Atmospheric Aerosol Particles, Fog Droplets, Cloud Droplets, Raindrops, and Snowflakes," *Rev. Geophys. Space Phys.* 21:903–920 (1983).

3. Bidleman, T.F. "Atmospheric Processes: Wet and Dry Deposition of Organic Compounds Are Controlled by Their Vapor-Particle Partitioning," *Environ. Sci. Technol.* 22:361–367 (1988).

4. Klöpffer, W., G. Rippen, and R. Frische. "Physicochemical Properties as Useful Tools for Predicting the Environmental Fate of Organic Chemicals," *Ecotoxicol. Environ. Saf.* 6:294–301 (1982).

5. Umweltbundesamt, Ed. Draft OECD-Test Guideline on Phogochemical Oxidative Degradation in the Atmosphere, Berlin, 2nd rev. (1987).

6. Klöpffer, W., R. Frank, E.-G. Kohl, and F. Haag. "Quantitative Erfassung der photochemischen Transformationsprozesse in der Troposphäre," *Chem. Ztg.* 110:57–62 (1986).

7. Klöpffer, W. "Rapid Test for Simulation of Photo-Oxidative Degradation in the Gas Phase," W. Funke, J. König, W. Stöber, A.W. Klein, and F. Schmidt-Bleek. Eds., Proceedings of the International Workshop on the Assessment of Photochemical Degradation of Substances in the Environment, Berlin, 194–205 (1980).

8. Klöpffer, W., G. Kaufmann, and R. Frank. "Phototransformation of Air Pollutants: Rapid Test for the Determination of K_{OH}," *Z. Naturforsch.* 40a:686–692 (1985).

9. Dilling, W.L., S.J. Gonsior, G.V. Boggs, and C.G. Mendoza. Organic Photochemistry. 20. Relative Rate Measurements for Reactions of Organic Compounds with Hydroxyl Radicals in 1,1,2-Trichlorotrifluoroethane Solution — A New Method for Estimating Gas Phase Rate Constants for Reactions of Hydroxyl Radicals with Organic Compounds Difficult to Study in the Gas Phase," *Environ. Sci. Technol.* 22:1447–1453 (1988).

10. Volman, D.H. "Photochemical Gas Phase Reactions in the Hydrogen-Oxygen System," *Advances in Photochemistry*, Vol. 1 (New York: Interscience, 1963), pp. 43–82.

11. Baulch, D.L., R.A. Cox, P.J. Crutzen, R.F. Hampson, Jr., J.A. Kerr, J. Troe, and R.T. Watson. "Evaluated Kinetic and Photochemical Data for Atmospheric Chemistry. Supplement I CODATA Task Group on Chemical Kinetics," *J. Phys. Chem. Ref. Data* 11:327–496 (1982).

12. Klöpffer, W. "How to Improve the H_2O_2 Photolysis as an OH Source in Simulated Atmospheric Degradation Testing of Chemicals?, *European Photochem. Assoc. (EPA) Newsl.* 31:4–12 (1987).

13. *Gemlins Handbuch der Anorganischen Chemie*, 8. Auflage. System Nr. 3, Lieferung 7, Sauerstoff (Weinheim, Bergstr.: Verlag Chemie, 1966).

14. Weeks, J.L. and M.S. Matheson. "The Primary Quantum Yield of Hydrogen Peroxide Decomposition," *J. Am. Chem. Soc.* 78:1273–1278 (1956).

15. Baxendale, J.H. and J.A. Wilson. "The Photolysis of Hydrogen Peroxide at High Light Intensities," *Trans. Faraday Soc.* 53:344–356 (1956).

16. Cox, R.A. and D. Sheppard. "Reactions of OH Radicals with Gaseous Sulphur Compounds," *Nature* 284:330–331 (1980).

17. Bünau, G.V. and T. Wolff. *Photochemie: Grundlagen, Methoden, Anwendungen* (Weinheim: VCH Verlagsgesellschaft, 1987).

18. Berufsgenossenschaft der chemischen Industrie (Herausg.): Wasserstoffperoxid, Merkblatt M 009 7/84 (1984), 1–22.

19. DEGUSSA (Herausg.). Wasserstoffperoxid; Eigenschaften, Handhabung und Anwendung, Frankfurt am Main (20. Jh.) ca. 1986.

20. *Handbuch der Gefährlichen Güter,* Merkblatt 206a (Berlin: Springer, 1987).
21. Finlayson-Pitts, B.J. and J.N. Pitts, Jr. *Atmospheric Chemistry Fundamentals and Techniques* (New York: John Wiley & Sons, 1986).
22. Rippen, G. *Handbuch Umweltchemikalien. Stoffdaten — Prüfverfahren — Vorschriften,* 2nd ed. (Landsberg: Ecomed, Loseblattsammlung, Stand 1988).
23. Frank, R. and W. Klöpffer. "Spectral Solar Photon Irradiance in Central Europe and the Adjacent North Sea," *Chemosphere* 17:985–994 (1988).
24. Behnke, W., F. Nolting, and C. Zetzsch. "The Atmospheric Fate of Di(2-Ethylhexyl)Phthalate, Adsorbed on Various Metal Oxide Model Aerosols and on Coal Fly Ash," *J. Aerosol Sci.* 18:849–852 (1987).
25. Behnke, W., W. Holländer, W. Koch, F. Nolting, and C. Zetzsch. "A Smog Chamber for Studies of the Photochemical Degradation of Chemicals in the Presence of Aerosols," *Atmos. Environ.* 22:1113–1120 (1988).
26. Eitzer, B.D. and R.A. Hites. "Atmospheric Transport and Deposition of Polychlorinated Dibenzo-p-dioxins and Dibenzofurans," *Environ. Sci. Technol.* 23:1396–1401 (1989).
27. Mill, T., M. Rossi, D. McMillen, M. Coville, D. Leung, R. Spanggord, and Platz, R. Photolysis of Tetrachlorodioxin and PCBs under Atmospheric Conditions. Report by SRI International, Menlo Park to VERSAR, Inc., Springfield (September 10, 1987).
28. Atkinson, R. "Estimation of OH Radical Reaction Rate Constants and Atmospheric Lifetimes for Polychlorobiphenyls, Dibenzo-p-dioxins, and Dibenzofurans," *Environ. Sci. Technol.* 21:305–307 (1987).
29. Atkinson, R. "Atmospheric Lifetimes of Dibenzo-p-dioxins and Dibenzofurans," *Sci. Total Environ.* 104:17–33 (1991).

CHAPTER 24

Abiotic Degradation Pathways of Selected Pesticides in the Presence of Oxygen Species in Aqueous Solutions

A. Mamouni, M. Mansour, and P. Schmitt

ABSTRACT

The photodegradation of carbetamide, chloridazon, and metoxuron in water in the presence of ozone, hydrogen peroxide, and oxygenated aqueous suspensions of titanium dioxide has been investigated. Photolysis kinetics were determined using solutions irradiated with a laboratory light source >290 nm and Heraeus suntest apparatus. The time required for 90% destruction was correlated to the concentration of the pesticides. The effect of the H_2O_2 concentration on the rate of carbetamide oxidation was determined. Several photoproducts were isolated and identified by spectroscopic methods. The results suggest that the degradation pathways of these compounds in the presence UV H_2O_2 and UV TiO_2 are hydroxylations of the aromatic ring. UV ozonation rapidly photooxidized all pesticides. The openings of the aromatic ring were observed, which produces lower molecular weight carboxylic acids; and further photooxidation, which eventually converts the acids to CO_2, H_2O, HCl, and NH_3.

I. INTRODUCTION

The direct and indirect photolysis of environmental contaminants represent an important degradative pathway for certain pollutants in the aqueous environment.

0-87371-616-7/93/$0.00 + $.50

Here one must distinguish between direct photolysis, refers to the process initiated by direct absorption of light by chemicals; and indirect photolysis, refers to those reactions in which another material absorbs sunlight and initiates chemical reactions that transform the chemical.

Photolysis of chemicals in aquatic environment is affected by sunlight, suspended particles, and dissolved organics; and by generated singlet molecular oxygen,[1] free radicals,[2] peroxides, superoxide anions,[3] hydrogen peroxide,[4] and alkoxyl radical in natural waters, suggesting alternative mechanisms for indirect photolysis.

Degradation processes in the environment are characterized by the splitting of the herbicide molecule. They include nonbiological, chemical, and photochemical decomposition and biological degradation by organisms and microorganisms. Persistent substances present difficult problems during treatment periods, because they are biologically decomposed very slowly. The use of ultraviolet (UV) irradiation in the presence of H_2O_2, O_3, or TiO_2, in combination with indigenous or selected soil microorganisms, has been examined as a disposal mechanism for decontamination of certain aqueous waste waters. Pretreatment of 2,4,6 trinitrotoluene (TNT) by UV O_3 yielded a number of more polar products that were more readily biodegradable when added to soil in comparison to the parent material.[5]

The photolysis of H_2O_2, TiO_2, and O_3 in aqueous solution produces OH radicals, which are known to react with a large number of environmental chemicals.[6] Because of their high electronegativity, OH radicals are very reactive oxidizing species.

The purpose of the present study was to determine the aquatic photolytic rate of pesticides, to identify the photodegradation products, and to determine the reaction pathways in aqueous solutions by means of UV irradiation in presence of oxidizing agents.

II. MATERIALS AND METHODS

The pesticides were purchased from Aldrich and were used without further purification. Solvents were reagents and analytical grades and obtained from E. Merck (Germany). TiO_2 (25%) and H_2O_2 (33%) were obtained from Degussa and Fluka, respectively. Ozone was generated by a Fischer ozone generator, model 503, using oxygen feed.

A. Photolysis

Aqueous solutions of pesticide and H_2O_2 or TiO_2 were placed in a cylindrical vessel, into which a Philips HPK 125 W high pressure mercury vapor lamp with a Pyrex cooling housing to block wavelengths shorter than 290 nm was inserted.

H_2O_2 and TiO_2 were added to the aqueous organic solution just before it was transferred into the reactor. O_3 or O_2 was bubbled through a sintered glass joined at the bottom of the vessel. The reaction temperature was kept under these conditions at 20–25°C.

B. Chemical Analysis

A UV/visible spectrum was obtained for each of the pesticides using a SP8-100 Philips spectrometer. The rate of disappearance of the pesticide in reaction mixtures was monitored by a high-performance liquid chromatograph (HPLC) equipped with a Gilson UV-vis detector. A reverse-phase (RP) column, 25 cm long, 4.6 mm i.d., RP packed with nucleosil (RP 18 5 μm) was used for analysis. The mobile phase was methanol:water (60:40) with a flow rate of 1.0 mL/min. A control sample was kept in the dark by converting the flask with aluminum foil to confirm that a given product was due only to photochemical reaction.

C. Analyses of Photoproducts

Analytical metoxuron, chlorodazon, and carbetamide in the amount of 4.4, 6.7, and 8.4 mmol, respectively, were dissolved in 3 L of distilled water with addition of hydrogen peroxide (300 μL/L) or titanium dioxide (200 mg/L). After irradiation, the solution was lyophilized and the mixture was separated by flash column chromatography (silica gel 100 g) using a gradient 0–100% ethylacetate in hexane and with acetone, collecting 50 mL fractions each time. Fractions were analyzed by using TLC. The photoproduct zones were located using 254-nm light.

Photoproducts which appeared to contain extraneous substances or which appeared to be mixtures were subjected to silica gel thin-layer chromatography (TLC) (thicknesses 2 mm). The plates were developed with (4:1:1) hexane:ethylacetate:acetone. After drying and scraping the silica gel containing the photoproducts from the TLC plate was collected in a filter paper and eluted with acetone.

The solvent was removed by rotary evaporation. The photoproduct was then dissolved in the appropriate solvent for identification.

D. Identification of Photoproducts

Photoproducts were identified by mass spectrometry (MS), gas chromatography (GC-MS), nuclear magnetic resonance (NMR), and infrared (IR). The mass spectrometers utilized were an LKB 9000s and a GC-MS Hewlett Packard 5992[A]; the gas chromatograph was equipped with a 25-m capillary (5%) phenyl methyl silicon. NMR spectra were run on a CFT 20 spectrometer (80 MHz) of ^1H and ^{13}C. The IR spectrometer was a Perkin Elmer 577.

E. Pesticides

Carbetamide (D) — [N-ethyl-2(phenyl-carbamoyloxy propionamide)] is a selective preemergence and postemergence herbicide that is used for control of annual grass and some broad-leaved weeds in alfalfa, sugar beet, and rape. The herbicide inhibits cell division in young tissues of roots and shoots.[7] Carbetamide has a relatively short persistance in soil (2 months) and is very stable to photodegradation, because it shows no UV absorption over 290 nm. Its photoreactivity in different organic solvents by UV light (<290 nm) has been reported.[8]

Metoxuron — [3-(3-chloro-4-methoxyphenyl)-,1-dimethylurea] is a selective herbicide and used as a pre- and postemergence. In plants, the main metabolic reactions are N-dimethylation and hydrolysis of the area moiety.[9]

Chloridazon — [5-amino-4-chloro-2-phenylpyridazin-3-2-(H)-one] is effective against broad-leaved weeds, particularly for use on sugar beet and beet crops. It persists in sandy loam for about 140 days.[10]

III. RESULTS AND DISCUSSION

A. Photodegradation with H_2O_2

Photolysis of carbetamide in distilled water ($\lambda > 290$ nm) presents no degradation, but in the presence of H_2O_2 all of the herbicide was depleted within 3 hr. The pH of the reaction with 200-ppm solution and 5×10^{-3} mol/L H_2O_2 was measured; and after 3 hr the pH had fallen from 6.2 to 4.4, indicating the formation of acids (Figure 1).

One can see from Figure 2 the effect of the concentration of H_2O_2 on the degradation rate of carbetamide. The most effective concentrations of H_2O_2 were those with the molar ratio of H_2O_2: carbetamide between 0.126 and 126. Therefore, it seems clear that there is an optimum concentration range of H_2O_2 (<5 $\times 10^{-1}$ mol). An excess of H_2O_2 in the solution actually retards the carbetamide degradation.

Chloridazon is readily photodegradable in sunlight.[11] The rate of degradation increased in the presence of H_2O_2 or TiO_2 (Figure 3). The results are similar to those observed in the photodegradation of metoxuron (metioned next).

Aqueous metoxuron (200) ppm was mixed with H_2O_2 with a molar ratio of H_2O_2:metoxuron = 0.12. The degradation rate is accelerated, with the generation of hydrogen chloride as proved by a pH decrease and a copious precipitate of silver nitrate solution. The pH of the reaction mixture falls rapidly from 6.66 to 3.08, but after 30 min the decrease of pH was very slow, suggesting that metoxuron was no longer photolyzing to II (Figure 4).

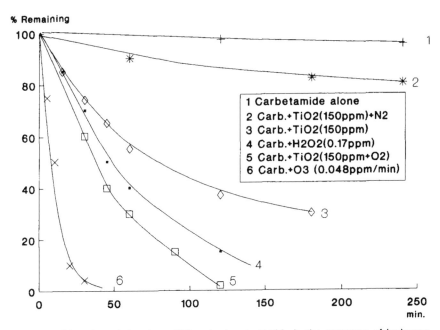

Figure 1. Photodegradation ($\lambda > 290$ nm) of carbetamide in the presence of hydrogen peroxide, titanium dioxide, and ozone in water.

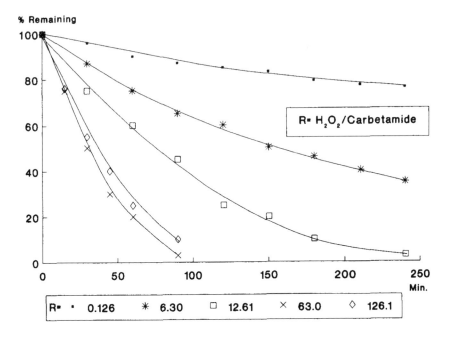

Figure 2. The effect of the concentration of H_2O_2 on the degradation rate of carbetamide.

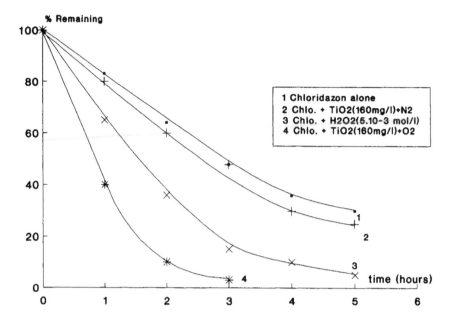

Figure 3. Photodegradation (λ > 290 nm) of chloridazon in the presence of hydrogen peroxide and titanium dioxide in water.

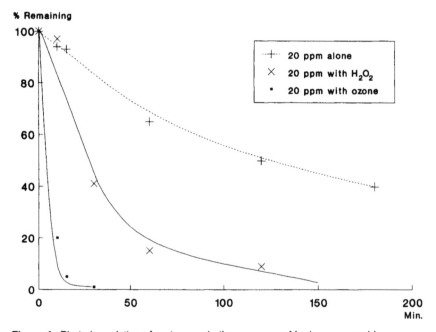

Figure 4. Photodegradation of metoxuron in the presence of hydrogen peroxide.

The primary reaction of the photodecomposition of hydrogen peroxide in an aqueous solution is considered to be a fission of the O–O bond of the excited molecule. It forms OH radicals, which initiate the chain decomposition reaction:

$$H_2O_2 + h\nu \rightarrow 2\ OH\cdot$$

$$OH\cdot + H_2O_2 \rightarrow H_2O + HO_2^\cdot$$

$$HO_2^\cdot + H_2O_2 \rightarrow H_2O + O_2 + OH\cdot$$

$$2\ HO_2^\cdot \rightarrow O_2 + H_2O_2$$

B. Photodegradation with TiO₂

With UV TiO₂ the degradation rate of all pesticides is hardly accelerated when oxygen is excluded. The half-life of carbetamide in the presence of TiO₂/oxygen was 0.6 hr. Under the same conditions but in the absence of oxygen, the extrapolated half-life increased to 40 hr indicating that the presence of oxygen is a necessity. Photodegradation curves for the loss of concentration of chloridazon are shown in Figure 3.

The generally accepted initial events in the irradiation of semiconductor TiO₂ dispersions are summarized by the following reactions:

$$TiO_2 + h\nu \rightarrow e^- + h^+$$

$$H_2O + h^+ \rightarrow OH\cdot + H^+$$

$$H^+ + e^- \rightarrow H\cdot$$

$$O_2 + e^- \rightarrow O_2^-\cdot \xrightarrow{\ H\cdot\ } HO_2^-$$

$$HO_2^- + h^+ \rightarrow HO_2^\cdot$$

$$2\ HO_2^\cdot \rightarrow O_2 + H_2O_2 \xrightarrow{\ O_2^-\ } OH\cdot + OH^- + O_2$$

(Huber-Weis reaction)

Substrate (RH) + OH· (or HO₂·) → ROH

Photoredox processes can occur at semiconductor interfaces as a result of the adsorption of "bandgap" radiation, preferably in the solar spectral region. The immediate products of this excitation are an electron in the conduction band and an electron vacancy or hole left in the valence band. Both these reactive species can migrate to the solid/solution interface and lead to change transfer across the interface and redox reactions in solutions.[15]

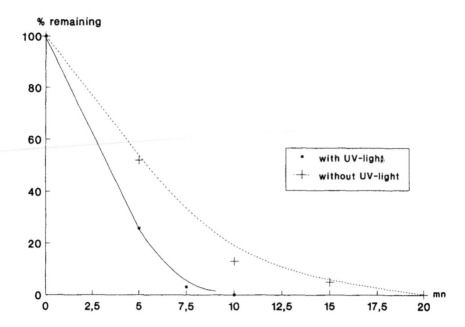

Figure 5. Degradation of chloridazon with ozone and UV light ($\lambda < 300$ nm).

C. Degradation with Ozone

The results show the rapid disappearance of all pesticides in the presence of this oxygen species, especially in the presence of ozone. The loss of chloridazon by ozonation alone, as compared to UV ozonation is shown in Figure 5. The rate of decomposition appears to be more rapid when compared to ozonation alone. This result is in agreement with Figure 6 which shows more consumption of ozone with UV ozonation. The concentration decreased by more than 90% in about 10 min and the pH decreased rapidly (Figure 7). The pesticide that photodegraded the slowest was carbetamide.

Ozone is a very weak oxidant. Direct ozonation reactions are highly selective, and only those compounds containing functional groups that are easily attacked by the electrophilic ozone can be oxidized. Ozone decomposition in aqueous solution leads to the formation of secondary oxidants, such as highly reactive radicals (OH$\cdot$, HO$_2\cdot$),[16] OH radicals[17] are the main secondary oxidants produced from decomposed ozone (Figures 8 and 9). UV decomposition of ozone in water leads primarily to the formation of hydrogen peroxide.[18] This hydrogen peroxide formed may decompose additional ozone into OH radicals.

IV. PHOTOPRODUCTS

The photometabolite NMR data and MS characteristic peaks are listed in Tables 1, 2, and 3.

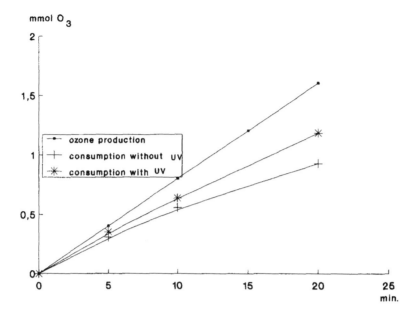

Figure 6. Ozone consumption on treatment of chloridazon in aqueous solution.

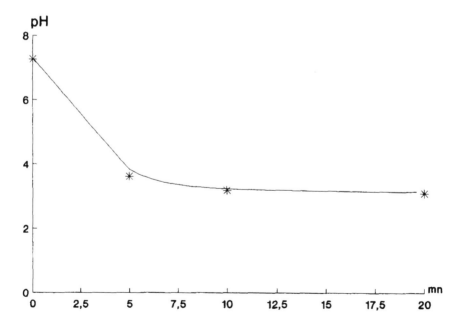

Figure 7. pH evolution during ozonation of chloridazon.

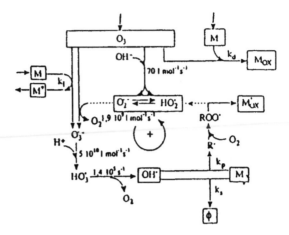

Figure 8. Reaction of ozonation in the presence of solutes M in water.

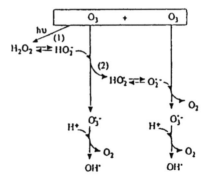

Figure 9. Pathways for transformation of ozone by UV photolysis.

A. Photoproducts of Carbetamide

Hydroxycarbetamide II was the major photoproduct isolated in solution with H_2O_2 or TiO_2, and it represented approximately 250 mg. The hydroxylation of carbetamide occurs at the ortho and para positions but not at the meta position. The ortho-hydroxylation product is preferential and appears first during the reaction, showing that the hydroxylation is the first step of the photodegradation.

Oxazoline dione IV (Figure 10) was isolated only in the presence of TiO_2. The cyclization may result from radical coupling with dissolved oxygen. From the degradation of the side chain, we have isolated phenyl-carbamoyloxy amino propionamide (50 mg) IV; it occurred by rupture of the C–N bonds of the second amide group. N-ethylactamide (70 mg) was present in the photolysate, but no aniline was detected or isolated. Aniline could be formed via phenyl isocyanate,

Table 1. Photometabolite NMR Data and MS Characteristic Peaks for Carbetamide

Structure and ^{13}C data	Substance	NMR^{1}H data	Mass data
	Carbetamide I Phenylcarbamoyloxy-2,4N-ethyl-propianamide	$-CH_3$ δ = 1.13 ppm J = 6.9 (T) $-CH_2$ δ = 2.50 ppm J = 6.9 (D) $-CH_2$ δ = 3.31 ppm J = 6.9 (M) $-CH$ δ = 5.21 ppm J = 6.9 (Q) NH δ = 6.7 Arom–H δ = 7.4 ppm (M)	236 (7%) M+ 191 (7%) 119 (100%) 93 (25%)
 II Hydroxy-phenylcarbamoyloxy-2N-ethyl-propionamide		$-CH_3$ δ = 1.13 ppm J = 7 (T) $-CH_2$ δ = 1.5 ppm J = 6.9 (D) $-CH_2$ δ = 3.31 ppm J = 6.9 (M) $-CH$ δ = 5.2 ppm J = 6.9 (Q) OH δ = 8.5 ppm NH δ = 6.6 ppm Arom–H δ = 7.7 & 6.9 ppm (M)	252 (10% M+ 207 (12%) 135 (100%) 119 (70%) 109 (30%) 91 (20%
 III Phenylcarbamoyloxy-amino-propionamide		$-CH_3$ δ = 1.13 ppm J = 7 (T) $-CH_2$ δ = 1.45 ppm J = 6.9 (D) $-CH_2$ δ = 3.31 ppm J = 6.9 (M) $-CH$ δ = 5.05 ppm J = 6.9 (Q) NH δ = 6.6 ppm Arom–H δ = 7.3 ppm (M)	208 (32%) M+ 119 (100%) 93 (70%)

Table 1 (continued). Photometabolite NMR Data and MS Characteristic Peaks for Carbetamide

Structure and ¹³C data	Substance	NMR¹H data	Mass data
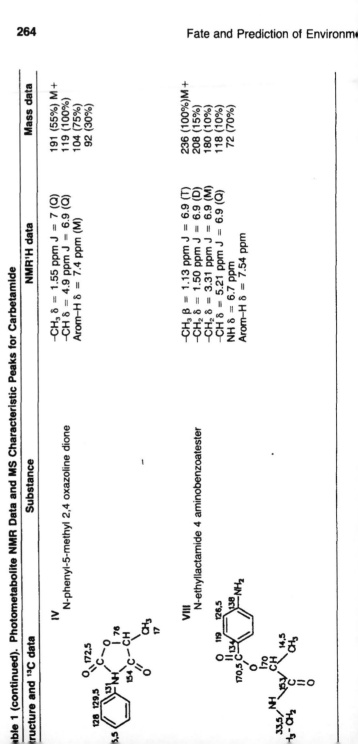	IV N-phenyl-5-methyl 2,4 oxazoline dione	$-CH_3\ \delta = 1.55\ ppm\ J = 7\ (Q)$ $-CH\ \delta = 4.9\ ppm\ J = 6.9\ (Q)$ Arom–H $\delta = 7.4\ ppm\ (M)$	191 (55%) M+ 119 (100%) 104 (75%) 92 (30%)
	VIII N-ethyllactamide 4 aminobenzoatester	$-CH_3\ \beta = 1.13\ ppm\ J = 6.9\ (T)$ $-CH_2\ \delta = 1.50\ ppm\ J = 6.9\ (D)$ $-CH_2\ \delta = 3.31\ ppm\ J = 6.9\ (M)$ $-CH\ \delta = 5.21\ ppm\ J = 6.9\ (Q)$ NH $\delta = 6.7\ ppm$ Arom–H $\delta = 7.54\ ppm$	236 (100%) M+ 208 (15%) 180 (10%) 118 (10%) 72 (70%)

Table 2. Photometabolite NMR Data and MS Characteristic Peaks for Chloridazon

Structure and ^{13}C data	Substances	NMR ^1H data	MS data
	Chloridazon I 5-Amino-4-chloro-2-phenyl-pyridazin-3-one	NH_2 δ = 6.9 ppm (S) Arom–H δ = 7.45 ppm (M) C–H δ = 7.75 ppm (S)	221 (57%) M+ 186 (5%) 158 (5%) 130 (7%) 105 (12%) 77 (100%)
II	5-N-(4-chloro-2-phenyl-pyridazin-3-one)-4'-N- (5'-amino-2'-phenyl-pyridazin-3'-one)-amine	N–H Δ = 6.0 ppm (S) NH_2 δ = 6.5 ppm (S) Arom–H δ = 7.4 ppm (M) C–H δ = 7.8 ppm (S)	406 (40%) M+ 390 (10%) 371 (38%) 299 (7%) 273 (5%) 236 (15%) 220 (40%) 145 (100%)
III	3,8-Diphenyl-dipyridazine-(2,7)-diol-pyrazine	Arom–H δ = 7.4 ppm (M) C–H δ = 7.8 ppm (S)	370 (100%) M+ 265 (12%) 237 (25%) 221 (50%)

Table 2 (continued). Photometabolite NMR Data and MS Characteristic Peaks for Chloridazon

Structure and ^{13}C data	Substances	NMR ^{1}H data	MS data
	IV 3,8-Diphenyl-dipyridazine-(2,7)-dione-pyrazine	Arom–H δ = 7.45 ppm (M) C–H δ = 7.70 ppm (S)	368 (45%) M + 331 (20%) 279 (85%) 221 (100%)

3. Photometabolite NMR Data and MS Characteristic Peaks for Metoxuran

structure	Substance	NMR ^{1}H data	Mass data
	**Metoxuron I** 3-(3-Chloro-4-methoxy-phenyl),-1-dimethylurea	CH_3 δ = 2.99 ppm (S) $O-CH_3$ δ = 3.85 ppm (S) Arom-H δ = 6.82 ppm J = 8.9 (D) Arom-H δ = 7.22 ppm J_1 = 7.9 J_2 = 2.6 (Q) Arom-H δ = 7.39 ppm J = 2.6 (D) NH δ = 6.39 ppm	228 (100%) M+ 214 (4%) 184 (68%) 168 (30%) 156 (10%)
	II 3-(4-Methoxy 2,5-paraquinone)-, 1-dimethylurea	CH_3 δ = 3.05 ppm (S) $O-CH_3$ δ = 3.85 ppm (S) Arom-H δ = 5.87 ppm (S) Arom-H δ = 7.26 ppm (S) NH δ = 7.84 ppm	224 (100%) M+ 209 (7%) 180 (53%) 166 (40%) 150 (7%)
	III (Hydroxymetoxuron) 3-(3-Hydroxy-4-methoxy-phenyl)-,-1-dimethylurea	CH_3 δ = 3.05 ppm (S) $O-CH_3$ δ = 3.82 ppm (S) Arom-H δ = 6.80 ppm J = 8.9 (S) Arom-H δ = 7.81 ppm J = 8.9 (S) NH δ = 6.69 ppm	244 (22%) M+ 228 (3%) 209 (100%) 180 (4%) 157 (5%)

(continued). Photometabolite NMR Data and MS Characteristic Peaks for Metoxuran

Substance	NMR ¹H data	Mass data
VI 3-(3-Chloro-4-methoxy phenyl)-,-1-methylurea O=C–NH N–C CH₃ H	CH₃ δ = 2.82 ppm J = 1.8 (D) O–CH₃ δ = 3.88 ppm (S) Arom–H δ = 6.87 ppm J = 8.6 (D) Arom–H δ = 7.18 ppm J₁ = 9 J₂ = 2.7 (Q) Arom–H δ = 7.31 ppm J = 2.6 (D)	214 (25%) M+ 183 (6%) 167 (50%) 142 (100%)
VII 3-(3-Chloro-4-methoxy phenyl)-,1-formyl-1- methylurea O=C–H O=C–N–CH₃ N H	CH₃ δ = 3.27 ppm (S) O–CH₃ δ = 3.88 ppm (S) Arom–H δ = 6.87 ppm J = 8 (D) Arom–H δ = 7.41 ppm J₁ = 8.8 J₂ = 2.5 (Q) Arom–H δ = 7.41 ppm J = 2.4 (D) CO–H δ = 8.5 ppm (S)	242 (15%) M+ 183 (100%) 168 (43%) 142 (14%)

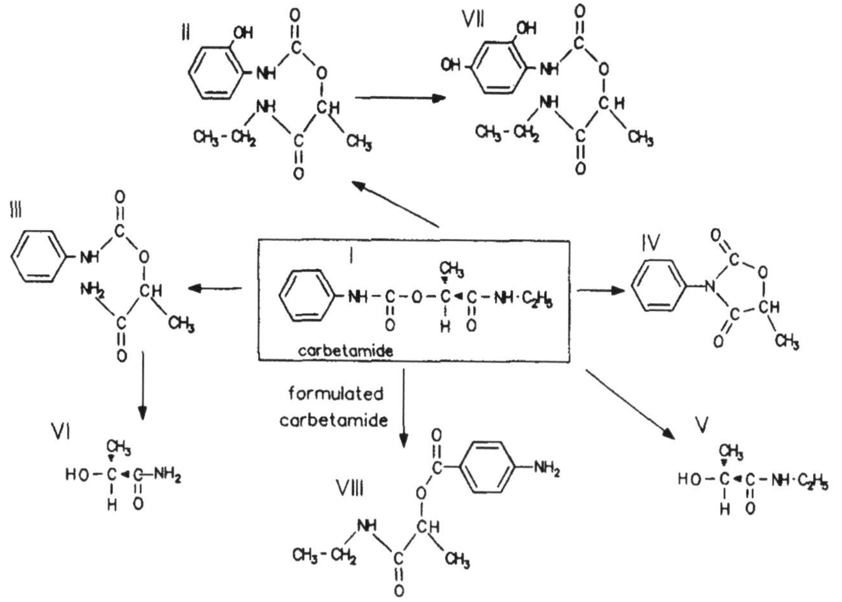

Figure 10. Photodegradation pathways of carbetamide in aqueous solution in the presence of hydrogen peroxide or titanium dioxide.

while isocyanates are unstable in water. Irradiation of formulated carbetamide was observed to form N-ethylactamide 4-aminobenzoate ester as the major photoproduct. This rearrangement is formally similar to the photo-Fries reaction of aryl esters.

B. Photoproducts of Chloridazon

An examination of the isolated photoproducts of chloridazon showed that, besides giving polymers products, it gave other products due to dechlorination as the first step in Figure 11. The formation of photoproduct II (30%) with its derived products, III (10%), and IV (39%) were the major metabolites after 2 hr of photodegradation in the presence of H_2O_2 or TiO_2. The complexity of NMR spectrum (^1H and ^{13}C) of photoproduct III can only be explained through the three mesomer forms in solution. The formation of photoproduct V, as a minor metabolite, can be visualized from I by photoaddition.

C. Photoproducts of Metoxuron

Photoproducts of metoxuron were identified by comparison of their spectra and a scheme for photodecomposition in aqueous solution is proposed in Figure 12. The quinone compound II [3-(4-methoxy 2,5-paraquinone)-,1-dimethylurea];

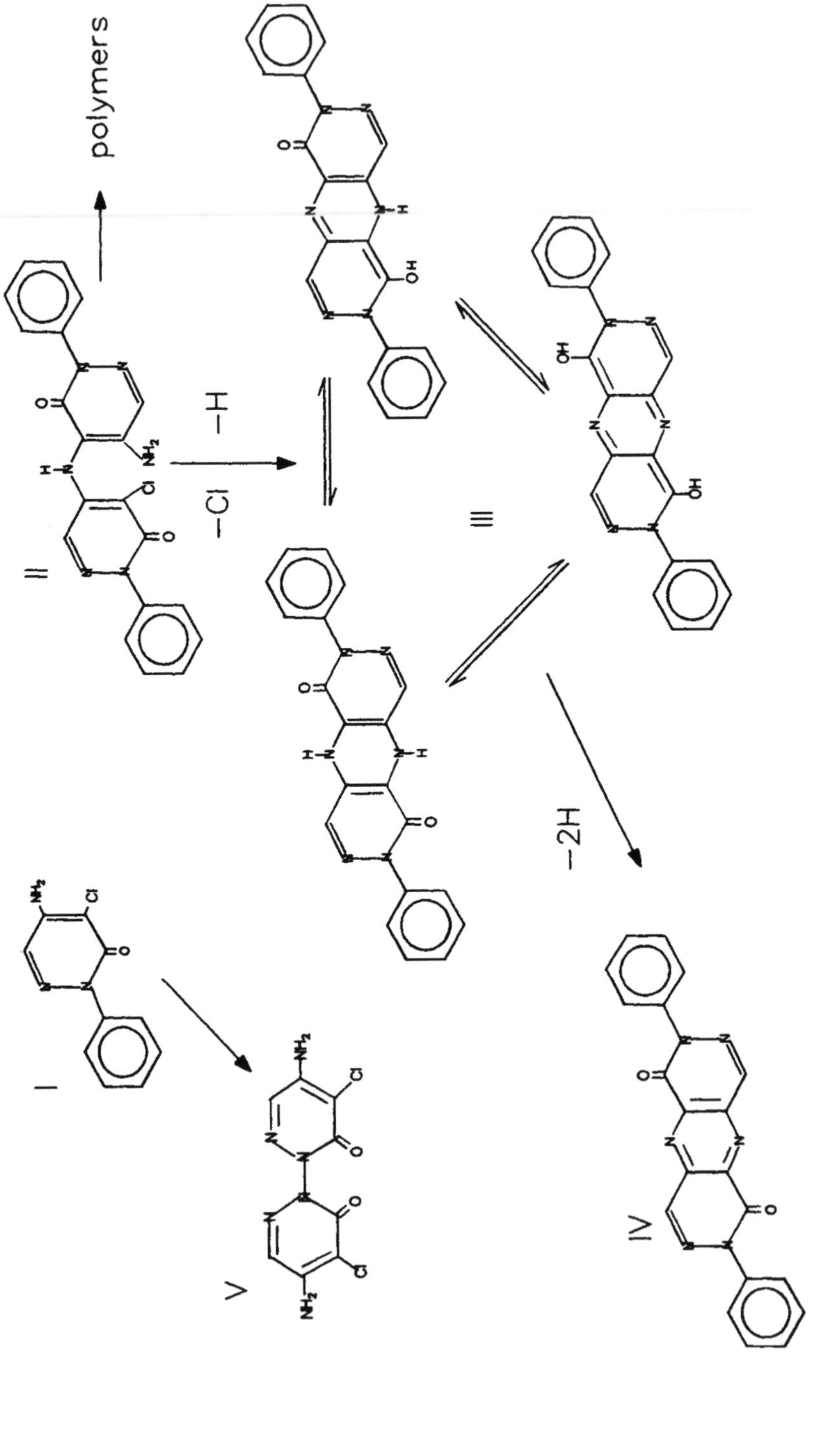

Figure 11. Possible alternative mechanisms for aqueous photolysis of chloridazon in the presence of hydrogen peroxide or titanium dioxide.

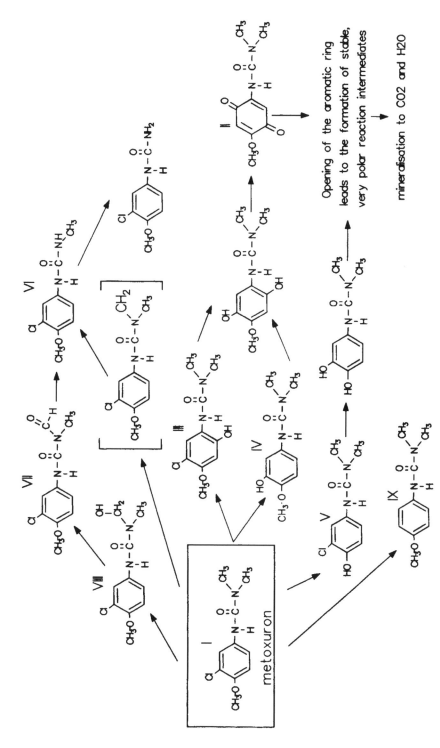

Figure 12. Proposed photodegradation pathways for metoxuron in aqueous solution in the presence of hydrogen peroxide or titanium dioxide.

hydroxy-metoxuron III, IV [3-(3-hydroxy-4-methoxyphenyl)-,1-dimethylurea], V [3(3-chloro-4-hydroxy phenyl)-,1-dimethylurea], and VI [3-(3-chloro-4-methoxyphenyl)-,1-methylurea] were the principal photoproducts. The first step of the reaction pathway (1) is removal of the chloro group from the benzene ring, replaced by a hydroxy radical group. Route (2) is the attack of OH radicals on the aromatic ring in position 6. A second hydroxylation of III and IV provides the quinone product II.

A large amount of VII [3-(3-chloro-4-methoxyphenyl)-1-formyl-1-methylurea] indicates that a part of the degradation is a side-chain oxidation. VIII [hydroxymethyl-metoxuron] produced the formyl product (the reverse reaction was not observed by irradiation of this formyl product). This compound was subsequently oxidized to acid. The resulting acid then appeared to undergo a decarboxylation to give product VI, while the acid was never detected in sufficient quantities for identification.

None of the photoproducts isolated in the presence of H_2O_2 or TiO_2 were observed by HPLC with carbetamide and chloridazon. With metoxuron only product VII was isolated as the minor product, and traces of II and III were observed and identified by HPLC with injection of the authentic sample.

We attempted to identify some of the mixture products by IR and NMR spectroscopy. The IR spectra clearly indicated the formation of carboxylic acids, which show a strong −COOH absorption band at about 1740 and 3500. The NMR spectra of herbicides in methanol-d_6 show weak nonresolvable proton resonances of the phenyl group at about 7 ppm, but the alkyl chain signals are still evident. This indicates that the aromatic ring has been decomposed, but that the aliphatic chain obtained is slowly decomposed.

The results suggest that the degradation pathways of these aromatic pesticides with ozone or UV/ozone occurred at the opening of the aromatic ring, which produces lower molecular weight carboxylic acids. Further photooxidation eventually rapidly converts the acids to full mineralization to CO_2, H_2O, and HNO_3. It explains that in the case of ozonation the hydroxylation products are only intermediate products. The isolation and identification of these final products have been proved to be complicated. The formation of CO_2 in the reaction was detected by formation of barium carbonate, $BaCO_3$, in barium hydroxide, $Ba(OH)_2$, during irradiation or ozonation.

It is also known that OH radicals are very reactive toward organic chemicals in water and the atmosphere. They possess a great affinity for electrons and react with a variety of chemical classes by abstraction of the hydrogen atom or by addition to a carbon-carbon double bond. The mechanism for the hydroxylation with OH radicals is the attack of an aromatic ring to form the hydroxycyclohexadienyl radical, which is eventually converted into hydroxy compounds. An electrophyllic character of the OH radical can explain why no meta-hydroxy compound was found in the present reactions.

V. CONCLUSION

Some chemicals, photochemically stable under environmental conditions or under other methods of degradation (such as chlorinated aromatic compounds), are in fact degradable in the presence of UV/O_3, UV/H_2O_2, and UV/TiO_2. The degradation of chlorinated organic substrates under the experimental conditions reported here is stable; and complete mineralization of these compounds gives CO_2 and HCl. These data show that this way could be used to detoxify wastewater, to degrade stable metabolites in which the microorganism could only catalyze the hydrolysis of the parent compound. It is probably not necessary for the photochemical treatment to completely decompose the organics, but it can be used to convert them to biodegradable compounds.

REFERENCES

1. Zafiriou, O.C. In *Chemical Oceanography, 2nd ed.*, Vol. 8, J.P. Riley and R. Chester, Eds. (London: Academic Press, 1983), pp. 339–379.
2. Mill, T., D.G. Hendry, and H. Richardson. *Science* 207:421 (1980).
3. Draper, W.H. and D.G. Crosby. Photochemical generation of superoxide radical anion in water. *J. Agric. Food Chem.* 31:734–737 (1983).
4. Draper, W.H. and D.G. Crosby. Photochemical generation of hydrogen peroxide in neutral water. *Arch. Environ. Contam. Toxicol.* 12:121–126 (1983).
5. Kearny, P.C., Q. Zeng, and J.M. Ruth. *Chemosphere* 12:1583 (1983).
6. Mansour, M. Photolysis of aromatic compounds in water in presence of hydrogen peroxides. *Bull. Environ. Contam. Toxicol.* 34:89–93 (1985).
7. Desmora, J., P. Ganter, P. Jacquet, and J. Metivier. *Phytiatr. Phytopharm.* 16:27–29 (1967).
8. Mansour, M., A. Mamouni, and P. Meallier. "Methodological Aspects of the Study of Pesticides Behaviour in Soil," P. Jamet, Ed. (Versailles: INRA, June 1989), pp. 89–100.
9. Worthing, C.R. *The Pesticide Manuel*, 8th ed.
10. Rosen, J.D. and M. Siewierski. *J. Agric. Food. Chem.* 2:(1972).
11. Anbar, M. and P. Neta. A complication of specific biomolecular rate constants for the reactions of hydrated electrons, hydrogen atoms and hydroxy radicals with inorganic and organic compounds in aqueous solution. *J. Appl. Radiat. Isotopes* 18:493–523 (1967).
12. Weeks, J.L. and M.S. Matheson. The primary quantum yield in the photolysis of hydrogen peroxide decomposition. *J. Am. Chem. Soc.* 78:1273–1278 (1956).
13. Adams, G.E., J.W. Boag, and B.D. Michael. Transient species produced in irradiated water and aqueous solution containing oxygen. *Proc. R. Soc.* 289(4):321–341 (1966).
14. Ikuichiro, I., W. Wendell, and K.O. Wilbourn. *J. Phys. Chem.* 84:3207–3210 (1980).
15. Bühler, R., J. Staehelin, and J. Hoigné. Ozone decomposition in water studied by pulse radiolysis. *J. Phys. Chem.* 88:2560–2564 (1984).

16. Hoigné, J. and H. Barder. Ozonation of water: oxidation competition value of OH radical reactions of different types of waters used in Switzerland. *Ozone Sci. Eng.* 1:357–372 (1979).

17. Hoigné, J. *Process Technologies for Water Treatment*, S. Stucki, Ed. (1988).

18. Hutzinger, O., Ed. *The Handbook of Environmental Chemistry* (Berlin: Springer-Verlag, 1982).

CONTRIBUTORS

Philippe Adrian, CNRS — Centre de Pedologie Biologique, F-54501 Vandouvre-Les-Nancy, France

B. Ahlsdorf, Federal Health Office, Institute for Water, Soil and Air Hygiene, D-1000 Berlin, Germany

Francis Andreux, CNRS — Centre de Pedologie Biologique, F-54501 Vandouvre-Les-Nancy, France

Robert A. Bulman, FJC — National Research Institute for Radiobiology and Radiohygiene, Department of Environment, H-1775 Budapest, Hungary

C. Catroux, Institut National de la Recherche Agronomique (INRA), Microbiologie du Sol, F-21034 Dijon, France

G. Celano, Instituto per lo Studio del Suolo, I-50121 Firenze, Italy

M.-P. Charnay, Institut National de la Recherche Agronomique (INRA), Microbiologie du Sol, F-21034 Dijon, France

Juan Cornejo, Instituto de Recursos Naturales y Agrobiologia, CSIC, Física y Química Ambiental, 41080 Sevilla, Spain

E. Diaz-Barrientos, Instituto de Recursos Naturales y Agrobiologia, CSIC, Física y Química Ambiental, 41080 Sevilla, Spain

U. Dörfler, GSF — Forschungszentrum fur Umwelt and Gesundheit GmbH, Institut fur Bodenokologie, D-8042 Neuherberg, Germany

J. C. Dur, Institut National de la Recherche Agronomique (INRA), Station de Science du Sol, F-78026 Versailles, France

C. Ehrig, Federal Health Office, Institute for Water, Soil and Air Hygiene, D-1000 Berlin, Germany

J.-C. Fournier, Institut National de la Recherche Agronomique (INRA), Microbiologie du Sol, F-21034 Dijon, France

Harald J. Geyer, GSF — Research Center for Environment and Health, Institute of Ecological Chemistry, W-8042 Neuherberg, Germany

Gunalan, Fakultas Pertanian, Universitas Sriwijaya, 30139 Palembang, Indonesia

María C. Hermosín, Instituto de Recursos Naturales y Agrobiologia, CSIC, Física y Química Ambiental, 41080 Sevilla, Spain

P. Jamet, Institut National de la Recherche Agronomique (INRA), Station de Phytopharmacie, F-78026 Versailles, France

A. A. Jukes, Horticulture Research International, Entomology Department, CV35 9EF Warrick, England

R. K. Khandal, Institut National de la Recherche Agronomique (INRA), Station de Science du Sol, F-78026 Versailles, France

W. Klöpffer, CAU GmbH, Battelle-Institut e.V., D-6000 Frankfurt am Main, Germany

E.-G. Kohl, CAU GmbH, Gesellschaft fur Consulting and Analytitc im Vmwelt bereich mbH, D-6000 Frankfurt am Main, Germany

Gerald Kuhnt, University of Kiel, Department of Geography, D-2300 Kiel, Germany

C. M. J. Lamaze, SUSDEVA, Environmental Management, CH-4125 Riemen, Switzerland

L. Madrid, Instituto de Recursos Naturales y Agrobiologia, CSIC, Física y Química Ambiental, 41012 Sevilla, Spain

A. Mamouni, GSF — Research Center for Environment and Health, Institute for Ecological Chemistry, D-8050 Fresing-Attaching, Germany

M. Mansour, GSF — Research Center for Environment and Health, Institute for Ecological Chemistry, D-8050 Fresing-Attaching, Germany

T. M. Miano, Universita di Bari, Istituto di Chimica Agraria, 70126 Bari, Italy

G. Milde, Federal Health Office, Institute for Water, Soil and Air Hygiene, D-1000 Berlin, Germany

E. Morillo, Instituto de Recursos Naturales y Agrobiologia, Física y Química Ambiental, 41080 Sevilla, Spain

Derek C. G. Muir, Freshwater Institute, Department of Fisheries and Oceans, Winnipeg Manitoba R3T 2N6, Canada

U. Müller-Wegener, Institute fur Wasser, D-1000 Berlin, Germany

B. Nickler, SUSDEVA, Environmental Management, CH-4125 Riemen, Switzerland

A. Piccolo, Instituto per lo Studio del Suolo, I-50121 Firenze, Italy

S. Lesley Prosser, FJC—National Research Institute for Radiobiology and Radiohygiene, Department of Environment, H-1775 Budapest, Hungary

Dominique Roche, Institut National de la Recherche Agronomique (INRA), Department of Phytopharmacy and Ecotoxicology, 78026 Versailles, France

Isabel Roldán, Instituto de Recursos Naturales y Agrobiologia, CSIC, Física y Química Ambiental, 41080 Sevilla, Spain

B. F. Rordorf, SUSDEVA, Environmental Management, CH-4125 Riemen, Switzerland

Irene Scheunert, GSF — Research Center for Environment and Health, Institute of Soil Ecology, D-8042 Neuherberg, Germany

Michel Schiavon, Ecole Nationale Superier d'Agronomie et des Industries Alimentaires, 54505 Vandouvre-Les-Nancy, France

R. Schmidt, Federal Health Office, Institute for Water, Soil and Air Hygiene, D-1000 Berlin, Germany

P. Schmitt, GSF — Research Center for Environment and Health, Institute for Ecological Chemistry, D-8050 Fresing-Attaching, Germany

P. Schneider, GSF — Research Center for Environment and Health, Institute of Soil Ecology, D-8042 Neuherberg, Germany

R. Schroll, GSF — Research Center for Environment and Health, Institute of Soil Ecology, D-8042 Neuherberg, Germany

N. Senesi, Universita di Bari, Istituto di Chimica Agraria, I-70126 Bari, Italy

D. L. Suett, Horticulture Research International, Entomology Department, CV35 9EF, Warrick, England

Gyula Szabó, FJC — National Research Institute for Radiobiology and Radiohygiene, Department of Environment, H-1775 Budapest, Hungary

M. Terce, Institut National de la Recherche Agronomique (INRA), Station de Science du Sol, F-78026 Versailles, France

S. Trapp, Mathematik/Informatik, University of Osnabrück, 4500 Osnabrück, Germany

R. Viswanathan, GSF — Research Center for Environment and Health, Institute of Soil Ecology, D-8042 Neuherberg, Germany

Stefan M. Waliszewski, University of Veracruz, Institute of Forensic Medicine, 91950 Veracruz, Mexico

Zhen-Hui Zhou, Academia Sinica, Institute of Entomology, Shanghai PRC, China

A. Zsolnay, GSF — Research Center for Environment and Health, Institute of Soil Ecology, D-8042 Neuherberg, Germany

INDEX

Printed and bound by CPI Group (UK) Ltd, Croydon, CR0 4YY

22/10/2024

01777605-0008